Anionic Polymerization: Principles and Practice

ANIONIC POLYMERIZATION: PRINCIPLES AND PRACTICE

Maurice Morton
Institute of Polymer Science
The University of Akron
Akron, Ohio

1983

ACADEMIC PRESS
A Subsidiary of Harcourt Brace Jovanovich, Publishers

New York London
Paris San Diego San Francisco São Paulo Sydney Tokyo Toronto

ACADEMIC PRESS, INC.
111 Fifth Avenue, New York, New York 10003

United Kingdom Edition published by
ACADEMIC PRESS, INC. (LONDON) LTD.
24/28 Oval Road, London NW1 7DX

Library of Congress Cataloging in Publication Data

Morton, Maurice.
Anionic polymerization: principles and practice.

Includes bibliographies and index.
1. Polymers and polymerization. I. Title.
QD381.M67 1983 547'.28 82-11627
ISBN 0-12-508080-8

PRINTED IN THE UNITED STATES OF AMERICA

83 84 85 86 9 8 7 6 5 4 3 2 1

To my wife
for her patience
and constant encouragement

Contents

Preface

Of all the mechanisms of polymerization reactions, the anionic mechanism is the *only* one which provides the chemist with an elegant tool for the synthesis of macromolecules having a remarkable *uniformity* of chain length, as well as a predicted molecular weight. This is, of course, an outcome of the stable nature of the growing chain, which can be prevented from undergoing any termination or transfer reactions. This characteristic feature of the anionic mechanism can also be utilized in the synthesis of block copolymers, branched polymers, and terminally functional polymers, all of which can be controlled to give desired molecular weights and molecular weight distributions.

Although the anionic mechanism is usually discussed along with the other mechanisms in books dealing with polymer chemistry, these unusual features and possibilities are not given the emphasis they deserve. Hence it is the objective of this book first to provide the reader with the necessary background for an understanding of the nature of the anionic mechanism of polymerization, and then to describe detailed aspects of the synthetic possiblilities inherent in these systems; the book has actually been divided into two parts in line with this treatment. In addition, an up-to-date review of the work done in these two areas is included.

It is expected that this book should be of special interest and real value to the large number of polymer chemists who work in the chemical industry, and who would wish to utilize the unique features of anionic polymerization in the synthesis of new products. It would also, of course, provide any chemist, whether professional or graduate student, with a definitive background in this field of polymer chemistry. It is hoped that the easily readable and concise style will make this book especially suitable for individual study, while at the same time providing the necessary literature sources for additional information.

The author wishes to express his appreciation and heartfelt thanks to Cheong Kow for his inestimable help in the monumental task of searching the literature for and collating the data contained in Chapters 2 and 8–11. He especially wishes to acknowledge, with deep gratitude, the dedicated and painstaking assistance of Barbara Zimmerman in preparing and editing the manuscript for this book.

Chapter

1 Introduction

1. Classification of Polymerization Reactions

Historically polymers have been divided into two broad classes: *condensation* polymers and *addition* polymers. Carothers[1] defines these as follows[2] (p. 37)*:

> ... *condensation* polymers, in which the molecular formula of the structural unit (or units) lacks certain atoms present in the monomer from which it is formed, or to which it may be degraded by chemical means, and *addition* polymers, in which the molecular formula of the structural unit (or units) is identical with that of the monomer from which the polymer was derived.

Thus an example of a condensation polymer would be a polyester, formed by the condensation reaction between a glycol and a carboxylic acid (with the evolution of water), whereas an addition polymer would be exemplified by polystyrene, formed by the self-addition of styrene monomers. It is obvious that these definitions are based on the *chain structure* of the polymer, regardless of its mode of formation. In fact, Flory[2] states (p. 40)*,

> It is thus appropriate to broaden the definition of condensation polymers to include not only those products which have been formed by a

* Reprinted from Paul J. Flory: "Principles of Polymer Chemistry." Copyright 1953 by Cornell University. Used by permission of Cornell University Press.

condensation polymerization process, as defined above, but also those polymers which on chemical degradation (e.g., by hydrolysis) yield monomeric end products differing in composition from the structural units.

Although this system of classification, based on structural considerations, has been very useful in the past, it has serious shortcomings in the light of present-day knowledge, especially in view of the variety of methods now available for polymer synthesis. Thus, for example, Nylon 6, a condensation polymer, is now synthesized by the ring-opening addition polymerization of caprolactam, and it is this process which is found to have a profound effect on the properties of the final polymer. This is, of course, due basically to the magnitude of the molecular weight of the polymer which results from the process. Since it is the extraordinarily large size of the polymer molecules which leads to their unusual properties, it behooves us to classify polymerization processes in accordance with *the way in which they affect the molecular size and size distribution of the final product*. On this basis, there appear to be two basic processes whereby macromolecules are synthesized, viz.

(i) Functional group polymerization,* and
(ii) Chain addition polymerization.

a. Functional Group Polymerization

These reactions fall into two classes:

(i) *Polycondensation*

$$A-R-A+B-R'-B \rightarrow A-R-R'-B+AB$$

(ii) *Polyaddition*

$$A-R-A+B-R'-B \rightarrow A-R-AB-R'-B$$

Here A and B are the functional groups which react with each other. Examples of polycondensation are seen in the formation of polyesters and polyamides, where the A and B groups would be hydroxyl and hydrogen, which would combine and split off as water. An excellent example of a polyaddition reaction of the above type is the reaction of diisocyanates with polyglycols or with hydroxyl-terminated polyesters, leading to polyurethanes. In this case no side products are formed.

The above polymerizations actually represent well-known reactions of small molecules, the only distinction being the minimum requirement of *difunctionality*, which makes it possible for the *product of each reaction to*

* This process has also been called *step-growth polymerization*, a somewhat confusing nomenclature, since *all* polymerization reactions involve "steps."

participate in further reactions. As a rule, the functional groups maintain their reactivity regardless of the chain length,* so that these reactions follow the same kinetic rules as simple molecules. As compared to polyaddition reactions, polycondensations suffer from the serious possibility of a reverse reaction (i.e., hydrolysis or "depolymerization") due to the accumulation of the by-product, e.g., water, and this must be taken into account.

The two factors which govern the molecular weight of the polymer in these polymerizations are

(i) stoichiometry, and
(ii) extent of reaction.

It is obvious that an excess of one type of end group will control the maximum chain length attainable, and this can be predicted if the initial ratio of functional groups is known. On the other hand, with equivalent amounts of the two end groups, the final chain length is theoretically limitless, i.e., infinite in size. In both cases, however, the chain length increases with extent of reaction, i.e., with time of reaction, since any molecule having the necessary functional group can react with any other molecule.

In the normal case, where the number of A and B groups are equal, the chain lengths are easily predictable as a function of the extent of reaction in the case of strictly linear polymers. It is obvious, however, that the actual chain lengths attainable depend on the extent of reaction attainable and that this factor poses a serious limitation on the final molecular weight. Thus even when such a reaction proceeds to 99% completion (a very high "yield" in terms of organic reactions), this results in polymer chains having an average of only 100 units per chain, i.e., a molecular weight of about 10,000. This is not considered a really high molecular weight for a polymeric material.

Since this type of polymerization is a completely random process, with all molecules having equal probability of reacting, the *distribution of molecular weights* corresponds to the *most probable* or *binomial distribution.*†

b. Chain Addition Polymerization

This type of reaction involves the successive addition of monomers to a growing chain, initiated by some reactive species. The addition reaction may involve either multiple bonds or ring opening. The reactive species which initiates the chain reaction must be capable of rupturing one of the bonds in the monomer, and may be either a radical, electrophile, or nucleophile. Hence the

* See p. 69 of Flory.[2]
† See p. 318 of Flory.[2]

three mechanisms* which operate in these polymerizations can be classified as (i) free radical, (ii) cationic, or (iii) anionic. Regardless of the mechanism, however, these reactions all share the characteristics of a *chain* reaction, i.e., each individual chain (both of the kinetic and physical variety) is started by an *initiation step*, followed by a series of propagation (or chain growth) steps. Hence the chain length, or molecular weight, need not be related to the extent of conversion of monomer to polymer. For example, very high molecular weight species may be formed even at very low conversion. Furthermore, the only reacting species, at any time, are the *growing chains* and the monomer, any interactions between polymer molecules being considered as "side" reactions.

These chain reactions may be very rapid, as in the case of the free radical or cationic mechanisms, where thousands of monomer units may add to a growing chain within a few seconds. On the other hand, they may be quite slow, as in the case of some of the anionic systems, where it may take hours for a growing chain to attain a high molecular weight. Nevertheless, all of these represent chain reactions, in that each chain molecule that is formed depends on an initiation step followed by a series of propagation steps.

2. Origins of the Anionic Polymerization Mechanism

It is usual for conceptual thought to have a slow and tortuous beginning but, once it is established, it appears simple and obvious. Thus we now readily accept the occurrence of the three mechanisms whereby unsaturated (or cyclic) monomers undergo addition polymerization, i.e., free radical, cationic, and anionic. Each of these is defined by the character of the growing *chain end*, i.e., whether it carries, respectively, an unpaired electron, a positive charge (real or formal), or a negative charge. However, the existence of these mechanisms was certainly not intuitively obvious, and their recognition was slow in coming.

Thus was it also for anionic polymerization systems, which were among the earliest to be used but among the last to be recognized. Because interest in synthetic rubber preceded Staudinger's macromolecular hypothesis by many decades (the first attempts to convert isoprene to rubber were recorded by Bouchardat in 1879 and by Tilden in 1882), it is not surprising that the use of sodium and potassium to polymerize isoprene dates back to the work by

* The recent rapid developments in stereospecific polymerization by means of complex organometallic catalysts (e.g., Ziegler–Natta systems) have led to such expressions as "coordinate anionic" in discussions of mechanisms. These are meant to refer to the complex organometallic species which result from the reaction of the initiators with the monomer and which then form the propagating chain end. However, since these species are still organometallic in nature, involving essentially a carbon–metal bond, they can still be considered as belonging to the anionic mechanism.

Matthews and Strange[3] in England in 1910, and by Harries[4] in Germany in 1911. Subsequently the polymerization of butadiene by means of metallic sodium became the prime method for synthetic rubber production in Germany and the USSR during the 1920s. The fact that sodium polybutadiene became the leading synthetic rubber can be ascribed mainly to its relatively high molecular weight, which could not be attained by other methods of polymerization. We can now understand this phenomenon in terms of the polymerization mechanism, sodium leading to an anionic mechanism involving little or no chain termination, while the "other" methods, e.g., heat, peroxide, or strong acids, result in either the free radical or cationic mechanisms, both of which involve relatively frequent chain termination or transfer steps.

Despite these industrial developments, the mechanisms of polymerization induced by alkali metals was only elucidated many years later. Thus Ziegler and his co-workers, in a series of papers[5-7] published in 1934, described the polymerization of butadiene, isoprene, 2,3-dimethylbutadiene, and piperylene by sodium, lithium, or alkyllithium species. This appears to be the first unambiguous report of the use of lithium or its derivatives in the polymerization of dienes. They considered all of these reactions as involving the "stepwise" addition of the diene to the organometallic species. Thus in the case of the reaction of sodium and butadiene, they postulated the initial formation of a disodium adduct as follows:

$$2Na + CH_2{=}CH-CH{=}CH_2 \rightarrow Na-CH_2-CH{=}CH-CH_2-Na$$

followed by subsequent stepwise addition of a diadduct to other butadiene molecules. They also showed that these reactions involved 1,2-addition as well as 1,4.

Ziegler,[8] at that time, actually proposed the presently accepted mechanism of polymerization of dienes and styrene by organolithium initiators, i.e., initiation by addition to the double bond, followed by consecutive addition of more monomer units to the growing chain. He even emphasized the absence of fortuitous termination of such chain reactions as long as the system was kept free of impurities, e.g., water, and thus actually defined what we presently accept as "living" polymers.

It was only in the post World War II period, however, that formal recognition of an anionic mechanism occurred. Thus Blomquist, Tapp, and Johnson[9] showed that the highly polar monomer nitropropene could be polymerized by alkaline potassium salts, while Beaman[10] showed that another polar monomer, methacrylonitrile, could be rapidly polymerized by the action of such strong bases as Grignard reagents and triphenylmethylsodium. He suggested that this polymerization proceeded by the nucleophilic attack of the growing carbanion on the electrophilic double bond of the monomer. Then Sanderson and Hauser[11] showed that even a nonpolar monomer such as styrene was polymerized by sodamide but only if liquid ammonia was used as

TABLE 1.1 Copolymerization of a 50/50 Mixture of Styrene and Methyl Methacrylate[12]

Initiator	Wt % styrene in copolymer
Benzoyl peroxide	51
Stannic chloride	99
Sodium	1
Potassium	1

solvent. They obtained only low molecular weights, presumably because of proton transfer from the ammonia solvent, and they also proposed a carbanionic mechanism, initiated by the amide anion.

It was also at that time that the clearest evidence for the existence of and distinction between the three possible mechanisms of chain addition polymerization was furnished by the work of Walling, Briggs, Cummings, and Mayo,[12] based on the copolymerization of styrene and methyl methacrylate. Their results are illustrated in Table 1.1. The profound differences in the copolymer composition were taken to indicate the existence of the three different mechanisms, i.e., the free radical peroxide initiator incorporating both monomers in the copolymer, while the cationic $SnCl_4$ highly favors the styrene and the anionic alkali metals are highly selective toward the electrophilic methacrylate monomer. This definitive investigation not only demonstrated the actions of the three mechanisms, but also illustrated the use of copolymerization as a powerful diagnostic tool.

3. Anionic Polymerization and Molecular Weights

As mentioned earlier, except for the observations of Ziegler,[8] the special features of anionic polymerization with regard to molecular weight were not recognized until fairly recently. This was mainly due to complications which were introduced by two features of the systems studied, i.e., heterogeneous reactions and side reactions with solvents. The first of these features was present in the very extensively used sodium polymerization of butadiene, and therefore obscured the molecular weight relationships. The second of these was present in such studies as that of Robertson and Marion[13] on the polymerization of butadiene by sodium in the presence of toluene, where very frequent chain transfer occurred with the toluene, leading to short-chain products, thus

$$\sim CH_2-CH=CH-CH_2Na^* + C_6H_5CH_3 \rightarrow \sim CH_2-CH=CH-CH_3 + C_6H_5CH_2Na \quad (1.1)$$

* The polybutadiene chain structure, which is shown here as 1,4, is actually a mixture, consisting largely of 1,2 units.

This reaction is defined as *metallation* of the toluene in the language of classical organic chemistry, but is considered as *chain transfer* by the polymer chemist.

A similar situation was demonstrated in the work of Higginson and Wooding[14] on the kinetics of polymerization of styrene by potassium amide in homogeneous solution in liquid ammonia. Because of the necessary use of the latter solvent, a high degree of chain transfer to the solvent was introduced as shown by the following proposed reaction scheme:

$$KNH_2 \rightleftharpoons K^+ + NH_2^- \tag{1.2}$$

$$NH_2^- + CH_2{=}CHC_6H_5 \rightarrow NH_2[CH_2\underset{\substack{|\\ C_6H_5}}{CH}]_xCH_2\underset{\substack{|\\ C_6H_5}}{CH^-} \tag{1.3}$$

$$NH_2[CH_3\underset{\substack{|\\ C_6H_5}}{CH}]_xCH_2\underset{\substack{|\\ C_6H_5}}{CH^-} + NH_3 \rightarrow NH_2[CH_2\underset{\substack{|\\ C_6H_5}}{CH}]_xCH_2\underset{\substack{|\\ C_6H_5}}{CH_2} + NH_2^- \tag{1.4}$$

It turned out, therefore, that reaction (1.4), which is a *chain transfer* step and *not* a termination step, strongly influenced the molecular weight of the product.

In the above case, the only true "termination" step would be the *reverse* reaction of (1.2), i.e., the reassociation of the *growing* dissociated ionic species, but this would not affect the molecular weight, provided this reaction was truly reversible. However, the chain transfer step (1.4) was apparently dominant in controlling the chain length, in this case anyway, thus overshadowing the absence of chain termination.

A similar kinetic analysis was later applied to the base-catalyzed polymerization of cyclic siloxane by Grubb and Osthoff,[15] who proposed the following scheme:

Initiation	$R_3SiOK \rightarrow R_3SiO^- + K^+$	(1.5)
Propagation	$R_3SiO^- + D_4 \rightleftharpoons R_3SiO[D]_x^-$	(1.6)
Termination	$R_3SiO[D]_x^- + K^+ \rightarrow R_3SiO[D]_xK$	(1.7)

where R_3SiOK represents a potassium silanolate initiator (which can be formed, for example, by the reaction of KOH with the cyclic monomer), and D_4 represents the cyclic monomer, in this case octamethylcyclotetrasiloxane, i.e., D represents the $(CH_3)_2SiO$ group.

It is obvious that reactions (1.5) and (1.7) simply represent the reversible nature of an ionic dissociation and that reaction (1.7) is therefore *not* a true, permanent termination step. In addition, of course, this polymerization is also distinguished by a *reversible* propagation step (reaction (1.6)), but this also should not affect the "living" nature of these growing chains.

Actually, kinetic studies of this system, both with the undiluted monomer[15] and in the presence of solvents,[16,17] confirmed the above kinetic scheme, e.g., the polymerization rate showed a half-order dependency on initiator concentration, a first-order dependency on monomer, and was also influenced by the dielectric constant of the solvent.[16] The latter would, of course, control the ionic dissociation equilibrium expressed by reactions (1.5) and (1.7) and thus control the concentration of growing chains. Unfortunately, this system was not a good one to illustrate the concept of the living polymer chain from molecular weight data, because of two special phenomena. In the first place, the stoichiometry which predicts that each initiator molecule generates a chain was obscured in the case of polymerization of undiluted monomer[15] by side reactions which apparently occur at the elevated temperatures ($\sim$ 150°C) required. On the other hand, the solution polymerization studies[16,17] at lower temperatures with the use of potassium methoxide instead of KOH did lead to the expected molecular weights.

However, even though the correct stoichiometry in molecular weights could be demonstrated, this system still lacked the other important feature of living polymerization, i.e., the narrow molecular weight distribution of the Poisson type. This was due to the fact that siloxane *bond interchange* is possible between the potassium silanolate chain ends and the in-chain siloxane bonds, leading to a reshuffling of polymer chain lengths, and hence to the usual most probable distribution.

It should be mentioned in this connection that the possibility of a Poisson distribution of molecular weights in the case of termination-free chain addition polymerization was recognized initially by Flory.[18,*] He treated the well-known case of the homogeneous polymerization of ethylene oxide by alkali alkoxides and pointed out that this type of chain growth is presumably termination-free in the absence of impurities capable of reacting with the alkoxide chain ends. Under such conditions, and if all chains have equal probability of growth, i.e., the initiation step is either similar in velocity or faster than the propagation step, the molecular weight distribution assumes the Poisson form, i.e.,

$$\begin{aligned} x_w/x_n &= 1 + (x_n - 1)/x_n^2 \\ &\sim 1 + x_n^{-1} \end{aligned} \tag{1.8}$$

It can readily be seen that this type of distribution "narrows" as the chain length (x_n) increases, and that it is, in principle, already remarkably narrow when $x_n = 100$ (i.e., $x_w/x_n = 1.01$).

Finally, the interest in anionic polymerization increased markedly as a result of two discoveries which occurred almost simultaneously. The first was the announcement[20] that lithium and its organic compounds could initiate the

* The general kinetics of termination-free addition polymerization was considered prior to that time by Dostal and Mark[19] but they did not consider the effect on the molecular weight distribution.

polymerization of isoprene to a very high (~ 90%) content of the cis-1,4 isomer, thus virtually duplicating the chain structure of the natural rubber molecule. It had been known[21] for some time that the isomeric chain structure of polybutadiene prepared by the use of alkali metals (e.g., Na, K) was influenced by the type of metal used, but the specific effect of lithium on the structure of polyisoprene was only elucidated with the advent of infrared spectroscopy as an analytical tool.

The second discovery was the finding[22] that nonterminating chain addition polymerization was not restricted to special systems, such as the ethylene oxide case described above, but was also possible for vinyl polymers. Thus Szwarc *et al.*[22] demonstrated that the homogeneous anionic polymerization of styrene initiated by sodium naphthalene solutions led to living polymers,* i.e., chains which showed no tendency to terminate their growth as long as any monomer remained. This discovery, which arose as a direct consequence of the very rapid initiation step in these homogeneous systems, was soon extended to other homogeneous anionic polymerizations involving not only styrene but other monomers, especially the dienes, which play a leading role in synthetic rubbers.

4. Special Features of Homogeneous Anionic Polymerization

Since the outstanding feature of homogeneous anionic polymerization is the absence of any fortuitous chain termination step, this enables these systems to accomplish the following control in the synthesis of polymers:

(i) synthesis of polymers having a predictable molecular weight average, from simple stoichiometry;

(ii) very narrow molecular weight distributions, approaching the Poisson distribution, by proper adjustment of initiation versus progagation kinetics;

(iii) synthesis of true block copolymers by sequential addition of different monomers to the living polymer chains; and

(iv) formation of polymers with functional end groups by selective termination with appropriate reagents.

It should be noted that the polymers described in (iii) and (iv) above would also have the features described in (i) and (ii). Hence it is no wonder that these anionic systems now provide the polymer chemist with the most elegant tool available for the synthesis of "tailor-made" homopolymers and block copolymers. It is especially noteworthy that such systems are capable of

* This somewhat euphemistic term was introduced[22] at that time and has been popularly adopted.

producing directly polymers of narrower molecular weight distribution than was ever possible by fractional separation from ordinary distributed polymers.

Because of the features described above, the subsequent chapters have been arranged in two parts: the first dealing with the nature and mechanism of these polymerization systems, and the second part describing their unique synthetic capabilities.

References

1. W. H. Carothers, *J. Am. Chem. Soc.* **51**, 2548 (1929).
2. P. J. Flory, "Principles of Polymer Chemistry." Cornell Univ. Press, Ithaca, New York, 1953.
3. F. E. Matthews and E. H. Strange, British Patent 24, 790 (1910).
4. C. H. Harries, *Justus Liebigs Ann. Chem.* **383**, 157 (1911); U. S. Patent 1,058,056 (1913) (to Bayer).
5. K. Ziegler, F. Dersch, and H. Willthan, *Justus Liebigs Ann. Chem.* **511**, 13 (1934).
6. K. Ziegler and L. Jacob, *Justus Liebigs Ann. Chem.* **511**, 45 (1934).
7. K. Ziegler, L. Jakob, H. Willthan, and A. Wenz, *Justus Liebigs Ann. Chem.* **511**, 64 (1934).
8. K. Ziegler, *Angew. Chem.* **49**, 499 (1936).
9. A. T. Blomquist, W. J. Tapp, and J. R. Johnson, *J. Am. Chem. Soc.* **67**, 1519 (1945).
10. R. G. Beaman, *J. Am. Chem. Soc.* **70**, 3115 (1948).
11. J. J. Sanderson and C. R. Hauser, *J. Am. Chem. Soc.* **71**, 1595 (1949).
12. C. Walling, E. R. Briggs, W. Cummings, and F. R. Mayo, *J. Am. Chem. Soc.* **72**, 48 (1950).
13. R. E. Robertson and L. Marion, *Can. J. Res., Sect. B* **26**, 657 (1948).
14. W. C. E. Higginson and N. S. Wooding, *J. Chem. Soc.* p. 760 (1952).
15. W. T. Grubb and R. C. Osthoff, *J. Am. Chem. Soc.* **77**, 1405 (1955).
16. M. Morton, M. A. Deisz, and E. E. Bostick, *J. Polym. Sci., Part A* **2**, 513 (1964).
17. M. Morton and E. E. Bostick, *J. Polym. Sci., Part A* **2**, 523 (1964).
18. P. J. Flory, *J. Am. Chem. Soc.* **62**, 1561 (1940).
19. H. Dostal and H. Mark, *Z. Phys. Chem., Abt. B* **29**, 299 (1935).
20. F. W. Stavely and co-workers, *Ind. Eng. Chem.* **48**, 778 (1956).
21. A. W. Meyer, R. R. Hampton, and J. A. Davison, *J. Am. Chem. Soc.* **74**, 2294 (1952).
22. M. Szwarc, M. Levy, and R. Milkovich, *J. Am. Chem. Soc.* **78**, 2656 (1956).

Part I

The Nature and Mechanism of Anionic Polymerization

Chapter

2 Monomers, Initiators, and Solvents

It is generally accepted that the anionic mechanism applies to those chain addition polymerizations in which the growing chain end has a negative charge (real or formal), and that these are initiated by bases of varying base strength. The latter, therefore, will generally contain a metallic cation, which becomes the counterion of the growing chain anion. It is obvious, therefore, that such a classification would encompass all organometallic initiators, including the vast array of transition metal complex initiators classified as the Ziegler–Natta systems. For the sake of clarity and simplification, however, as stated earlier, this discussion will be limited to the simple anionic systems where the initiators, heterogeneous or homogeneous, are generally, but not exclusively, based on the alkali (or alkaline earth) metals.

The classes of monomers involved in anionic polymerization are of three main types: monomers based on the carbon–carbon double bond, heterocyclic monomers, and monomers based on the carbon–heteroatom double (or triple) bond. The first two of these have been found to be quite susceptible to base-catalyzed polymerization (e.g., dienes, cyclic oxides, cyclic sulfides), while the latter (e.g., aldehydes) are not so clearly involved. As might be expected, these different classes of monomers require initiators of differing basicity. Furthermore, the active chain ends in these systems exhibit different reactivities, e.g., carbanionic chain ends being much more reactive than alkoxide chain ends toward water and other atmospheric components.

Because of the discovery of the stereospecific polymerization of isoprene by lithium-based catalysts, as well as of the living nature of many anionic polymerizations, there has been a great deal of interest in these systems over the past twenty years. This has spurred the publication of a number of reviews of this field, dealing mainly with the nature and mechanism of these reactions.[1-9] One of these[7] is of special interest in that it attempts to collate available data on the types of monomers, initiators, and solvents used in anionic systems. Although this type of information suffers from gradual obsolescence, it is still very useful to have it in one place, and has therefore been updated and reproduced herein, in Tables 2.1, 2.2, and 2.3, which deal, respectively, with vinyl monomers, monomers containing carbon–heteroatom multiple bonds, and cyclic monomers.

Although it would be desirable to evaluate these data critically, this would not only involve a lengthy and tedious task, but would probably lead to doubtful conclusions, mainly because of the difficulty of evaluating the experimental techniques used. Since carbanions especially are known to be very sensitive to the presence of protic solvents or other reactive impurities, the only reliable data obtained in such polymerizations are from experiments carried out under rigorous conditions. The type of precautions needed and the special high vacuum techniques involved have been described recently.[9]

In view of these difficulties involved in any attempt to evaluate available data, it was decided to limit this review to a simple classification and tabulation of the various monomers, initiators, and solvents reported by various investigators, including the *reported* results, without comment. The reader is thereby at least referred to the source of the quoted information.

TABLE 2.1 Monomers Based on Carbon–Carbon Double (and Triple) Bonds

Monomer	Initiator system	Remarks	References
Styrene and Derivatives			
Styrene	Na naphthalene in THF	Stability of carbanion demonstrated	10–35
	Na biphenyl	Monodisperse polymer	36
	Na anthracene	Relatively slow initiation; anthracene complexes with active chain end	37–39
	Na naphthalene in *p*-dioxane	High molecular weight	40–42
	α-Phenylethylpotassium in THF	Monodisperse polymer	43
	Organolithiums in hydrocarbons and ethers	Termination free, monodisperse polymer	44–64
	Butyllithium in toluene at low temperatures	Crystallizable polymer; may be due to presence of alkali hydroxides and other salts; see Ref. 61	65–68
	Triphenylpotassium in toluene		69
	Potassium in liquid NH_3	Initiation due to free NH_2 ion; low molecular weight	70–72
	Sodium in liquid NH_3	Initiation due to free NH_2 ion; low molecular weight	73
	Na, K, Rb, and Cs counterions in benzene	Some termination for Na; other counterions form dimers and are termination free	74
	Na naphthalene in THF	Isomerization reaction occurs following completion of the polymerization; termination results	75, 76
	Sodium in ether	High polymer	77
	1,2-dilithio-1,2-diphenylhexane and 1,2-dilithio-1,2,3,4-tetraphenyloctane in benzene	Monodisperse ($M_w/M_n \simeq 1.1$) polystyrene; dilithium initiators are soluble in benzene	78
	BuLi + Na hexylate 45–65°C	Faster reaction rate than BuLi alone; broader MWD ($M_w/M_n = 1.35$–1.95)	79
	$NaAlH_2(OCH_2CH_2OCH_3)_2$, $NaB\Phi_4$	Polystyrene of $M_w/M_n = 3.1$ and 1.08, respectively, were obtained	80
	Polystyryllithium in benzene	Dioxane affects the polymerization mechanism	81
	Na and Ca in THF or 1,2-dimethoxyethane	The respective activation parameters determined for both the ion pair and the free ion	82

(*table continues*)

TABLE 2.1 Double and Triple Bond Monomers (*continued*)

Monomer	Initiator system	Remarks	References
Styrene (*continued*)	Na naphthalene in THF	The initiator system gave at least 2 species which showed completely different behavior in initiation	83
	Polystyrylsodium	Mechanism studies	84
	Li in benzene–dimethoxyethane mixture at 25°C	Electric field effect on the polymerization investigated	85
	BuLi in toluene at 20°C	Effect of anisole, *o*-ethylanisole, and *p*-ethylanisole on the rate constants	86
	Li, Na, K, and Cs	Electric field effect	87
	RMgX in $(Me_2N)_3PO$	High polymer obtained when $[(Me_2N)_3PO]/[RMgX] > 2$	88
	Amylsodium in ethylbenzene with $[Si(CH_3)_2O]_4$ as additive	Polysiloxane increases rate and forms polystyrene of high MW and narrow MWD ($M_w/M_n = 1.58$)	89
	Na naphthalene	Dicarbanion salt reduced with a halogen yielded polymer having predominantly head-to-head and tail-to-tail linkages	90
	Dimsylsodium	Low MW polymer ($\leqslant 20{,}000$)	91
	s-BuLi in cyclohexane or benzene	Degree of association determined	92
	Sodium thiophenoxide in polar aprotic solvents, e.g., Me_2SO, $(Me_2N)_3PO$	Living polymer; no polymerization initiated by the initiator in less polar solvent	93
	t-BuLi, *n*-BuLi, *s*-BuLi in aliphatic and aromatic hydrocarbon	Comparison of initiation rates	94
	$Ba(\Phi CH_2)_2$ in THF	Living polymer; polymerization retarded by presence of barium tetraphenylboride which suppressed ionization of living ends	95, 96
	s-BuLi in hexane	A 75:25 styrene–*t*-butylstyrene block copolymer was used as a dispersing agent in a slurry process	97
	Polystyrylsodium, -potassium, and -lithium in 2-methyltetrahydrofuran	Influence of an electric field on rate	98
	EtLi, Na α-methylstyrene tetramer in triglyme–benzene mixtures	Kinetics of polymerization	99

Dibenzylmagnesium/tetrahydrofurfuryl alcohol in THF at − 70°C	No polymerization induced with the magnesium alkyl alone	100
R_2Mg/2-alkoxyethanol	No polymerization induced with dialkylmagnesium alone	101
Difunctional organobarium in THF	For a given temperature the propagation rate does not depend on carbanion concentration (3×10^{-5} to 5×10^{-3} *M*)	102
Disodium α-methylstyrene tetramer in tetrahydropyran at 25°C	Electric field effect on the polymerization rates	103
Cumylcesium in dimethoxyethane + 25 to − 55°C	Kinetics of polymerization in presence of a common ion salt	104
n-BuLi in toluene	Chain transfer to toluene suggested	105
$NaB\Phi_4$ in THF	Kinetics of polymerization	106
Cumylsodium in cumyl methyl ether	Kinetics and MWD	107
BuLi in benzene/THF mixtures	Effect of electric field on the propagation rate and living end concentration	108
Cumylcesium in THF	Kinetics of polymerization	109
Na α-methylstyrene tetramer in $(Me_2N)_3PO$	Kinetics study	110
Na α-methylstyrene tetramer in polar solvents	Influence of polar solvents on ion/ion pair equilibrium	111
Alkali metals in THF	Polymerization in the presence of macroheterobicyclic	112
Diethylamide in benzene/THF mixtures	Effect of solvation and coordination process on initiation and propagation steps	113
Φ_3CCaCl, fluorenylcalcium chloride, diindenylcalcium	Effective initiators, but wide MWD obtained	114
Dilithiostilbene, disodio-α-methylstyrene tetramer in toluene	Chain transfer when Na^+ is the counterion	115
Na in Me_3N	Kinetics of polymerization $M_w/M_n = 1.24$ (MW ~ 200,000)	116
Cumylpotassium in THF at − 70°C	Kinetics of polymerization	117

(*table continues*)

TABLE 2.1 Double and Triple Bond Monomers (*continued*)

Monomer	Initiator system	Remarks	References
Styrene (*continued*)	Cumyl-Cs or Na-α-methylstyrene tetramer, in THF or 2-methyltetrahydrofuran	Influence of electric field on kinetics	118
	Polystyrylsodium in dioxane	Propagation rate constant determined	119
	Na α-methylstyrene tetramer in tetrahydropyran	Kinetics study	120
	Graphite–K inclusion compounds in THF, 1,2-dimethoxyethane, benzene, heptane at 23°C	Yield depends on the solvent employed	121
	EtLi in toluene	Polymerization rate and MW affected by addition of anthracene	122
	BuLi in mercury	Colorless polystyrene with MW 40,000	123
	t-BuMgCl, EtMgBr, ΦMgBr, BuMgCl, Bu_2Mg, $(Et_2N)_2Mg$ (in $Me_2N)_3PO$)	Atactic polymer	124
	BuLi or *s*-BuLi in toluene at − 50 to 25°C	Isotactic PS was formed at temperature below − 10°C	125
α-*n*-Alkylstyrenes (alkyl ≥ C_2H_5)	Na naphthalene in THF at − 75°C	Slow initiation; low MW	126
Halogenated styrene	Na naphthalene, Na biphenyl, Na α-methylstyrene tetramer, and BuLi	More than 80% living ends and can be functionalized	127
α-Methylstyrene	Na naphthalene in THF	Complete conversion at − 78°C	128–131
	Na in THF	Heterogeneous initiation at room temperature; polymerization carried out at − 78°C	132, 133
	n-BuLi in THF	Initiation at room temperature; polymerization carried out at − 78°C; monodisperse polymer	134
	Na naphthalene in THF	Formation of trimer and tetramer at temperatures of − 30 to 25°C	135, 136
	BuLi in bulk polymerization at room temperature	An equilibrium reaction exists and the MWD are broad (M_w/M_n = 1.55–2.2)	137

	Na naphthalene in THF (– 78°C), tetrahydropyran or hexane–THF mixtures	Dependence of tacticity on polymerization conditions	138
	BuLi in toluene at – 50°C	Reaction accelerated by hydroxyalkylated polyhydric phenols	139
	RMgBr in $(Me_2N)_3PO$	Reactivity changes with R in the order benzyl > allyl > aryl ~ *s*-alkyl > *n*-alkyl	140
	Na in THF/toluene mixture at – 25 to 5°C	Kinetics of polymerization	141
	Na, K in THF at – 30 to 10°C	Kinetics of polymerization	142
	Na^+, K^+, Rb^+, and Cs^+ at – 70°C in THF	Addition of triglyme forms the ether-separated ion pairs and increases the ion-pair propagation constant	143
	K in *p*-dioxane	The monomer–polymer equilibrium was studied	144
	Na naphthalene in bulk at – 30 to 50°C	Study of monomer–polymer equilibrium	145
Substituted α-methylstyrenes	Sodium dispersion in hexane or heptane	$(Me_2N)_3PO$ used as an accelerator; anionic dispersion system, with poly(vinyl ethyl ether) as dispersing agent	146
trans-β-Methylstyrene	BuLi in hydrocarbon solvent or THF	Chain transfer reaction due to active H of the monomer	147
o- and *p*-Methylstyrenes	Sodium in 2-methyltetrahydrofuran	The influence of the CH_3 group on the kinetic and dissociation behavior	148
	EtLi, cumyl-Cs, Na- or K-α-methylstyrene tetramer, in 2-methyl tetrahydrofuran	The influence of electric field on the kinetics	149
o- and *p*-Methoxystyrene	BuLi in toluene at 20°C	*o*-Methoxystyrene similar to styrene, while the *p*-isomer has slow initiation, relatively fast termination and forms a highly branched or crosslinked polymer	150
β-Nitrostyrene	MeONa in MeOH at 22°C	Crystalline polymer	151
	Sodium ethoxide in ethanol/THF	Kinetic studies; some narrow MWD	151a

(*table continues*)

TABLE 2.1 Double and Triple Bond Monomers (*continued*)

Monomer	Initiator system	Remarks	References
Acrylates			
Acrylate and methacrylate esters (stereospecifically deuterated)	Fluorenyllithium, BuLi, fluorenylcesium, $LiAlH_4$, $\Phi MgBr$, 1,1-diphenylhexyllithium in toluene, THF, and ΦCH_3–Et_2O, ΦCH_3–THF mixtures	Highly isotactic polymers under several conditions	152
Acrylates (α-substituted)	$\Phi MgBr$, BuLi	Increase in the isotacticity of the polymers formed upon addition of small amount of H_2O, HCl, or CH_3OH	152a
9-Acrylcarbazole	*n*-BuLi in toluene	High conversions (80–90%) below a temperature of 0°C	153
1-Acryloxy-2-butyne (and 1-methacryloxy-2-butyne) $CH_2{=}C(R){-}C({=}O){-}OCH{-}C{\equiv}CR'$ R = H or CH_3; R′ = H, lower alkyl	Li naphthalene, *n*-BuLi, Na naphthalene, Na benzalaniline, in THF or toluene	Soluble polymer with pendant acetylenic groups for sodium catalysts, but crosslinked polymers obtained with organolithium compounds	154
Acryloylverdazyl, methacryloylverdazyl 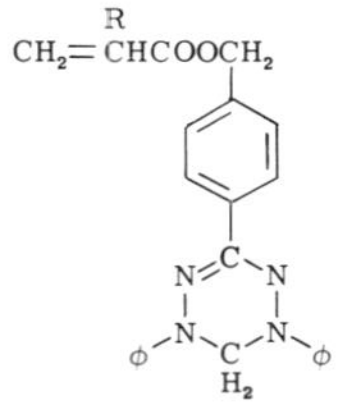	*n*-BuMgBr or Na naphthalene	Dark green polymer (MW ~ 2300) containing ~68% stable free radical	155

Alkyl α-chloroacrylates	n-BuLi, n-BuLi/EtOLi, 1,1-diphenylhexyllithium, t-BuLi, ΦLi, ΦMgBr, tritylsodium, ΦNa in toluene at low temperatures	Polymer tacticity depends on polymerization conditions	156
Methyl α-chloroacrylate	ΦMgBr, BuLi or $LiAlH_4$ in toluene	Polymer tacticity studied by NMR	157
	n-BuZn or butyl dimethylmalonate–Zn in $(Me_2N)_3PO$	Polymer insoluble in MeOH	158
	1:1 ΦMgCl–benzalacetophenone	Crystallizable polymer	159
	ΦMgBr, n-BuLi	Amorphous polymer	160
Ethyl α-chloroacrylate	n-BuLi, ΦMgBr; fluorenyllithium in THF or toluene at − 78° or 0°C	MW < 20,000 Low yield with Grignard reagent	161
Methyl (and ethyl) α-chlorothiolacrylate	n-BuLi, ΦMgBr	Crystalline polymer	160
Alkyl α-cyanoacrylates	$HCONMe_2$ in CH_3NO_2	Kinetics and activation energies	162
Isobutyl 2-cyanoacrylates	Amines	Gel polymers	163
Alkyl methacrylates	Li, Na, and K biphenyl in 1,2-dimethoxyethane and THF at − 70°C	Effect of cation, solvent, and ester on the stereoregularity of the poly(alkyl methacrylate)	164
Allyl acrylate and methacrylate	Na naphthalene in THF and n-BuLi in toluene	Linear, soluble polymer formed; allyl group does not participate in the polymerization	165
Aryl (and methyl) methacrylates	ΦMgBr, BuLi in toluene and THF at low temperatures	Stereoregularity of the polymers and the relative reactivity of the monomer	165a
Benzyl methacrylate	n-BuLi, ΦMgBr	Predominantly isotactic at lower temperatures (− 78°C). By increasing the polymerization temperature, the isotacticity decreases, but the heterotacticity increases, while the syndiotacticity remains almost constant	166
Bis(triethylene glycol) phthalate dimethacrylate	Na naphthalene	Equilibrium between soluble and crosslinked polymers in the living system; yield depended on catalyst concentration and showed a maximum	167–170
Butyl methacrylate	t-BuONa/BuOH	Polymer of MW 600–100,000	171
	MeONa/MeOH	Oligomers obtained	172
	t-BuOK, t-BuLi, $Φ_3CNa$, t-BuONa in benzene	Kinetics of polymerization	173

(*table continues*)

TABLE 2.1 Double and Triple Bond Monomers (*continued*)

Monomer	Initiator system	Remarks	References
2-(*N*,*N*-Dialkylamino)ethyl methacrylate	K or KOH at 60°C	Soluble polymer, thermally stable to 200°C	174
	t-BuOLi or *t*-BuLi	Polymer less isotactic than PMMA because of the presence of electron donor	175
(2,2-Dimethyl-1,3-dioxolan-4-yl)methyl methacrylate	Disodium 1,1,4,4-tetraphenyl-1,4-butane in THF at − 60°C	Linear polymers of low polydispersity	176
Diphenylmethyl methacrylate	*n*-BuLi, ΦMgBr	A highly isotactic polymer in toluene at low temperatures and a highly syndiotactic one in THF	166
1,1-Diphenylethyl methacrylate	*n*-BuLi in toluene or THF at − 78° to − 30°C	Stereospecific polymerization	177
Hydroxyalkyl acrylate	LiH, *t*-BuOLi, NaH, MeONa, *t*-BuOK, Φ_3P (DMF or THF at 50 to 80°C)	Good yield (43–93%); polymerization through hydrogen-transfer mechanism; polymer having structure $\left(CH_2CH_2-CO_2-R-O\right)_n$ prepared	178
9-*O*-Methacroylquinine	BuLi in toluene	Polymer has an asymmetric conformation in the main chain	179
2-Methoxyethyl methacrylate	*t*-BuOLi in nonpolar media	Two growth regions separated by a plateau were observed in the conversion curve; change of course of polymerization accompanied by changes in the microstructure of polymer	180
	t-BuOLi or *t*-BuLi	Polymer less isotactic than PMMA because of the presence of electron donor	175
Methyl acrylate	*n*-BuLi in hexane and THF	Side reactions between initiator and monomer studied	181
	n-BuLi in hexane and THF	*n*-BuLi is consumed by conjugate addition and *n*-butane formation	181
	Sodium salts of fluorene, anthracene, biphenyl, benzophenone, 1,1-diphenylethylene, naphthalene, tritan in THF at − 30 and − 75°C	Mode of action of each catalyst discussed	182

Methyl α-*n*-alkylacrylates (alkyl $\geq C_{12}H_{25}$)	Na naphthalene (– 75°C) or Na (0°C or room temperature) in THF or toluene	Soluble polymer	183
Methyl α-alkylacrylates (alkyl $> C_1; \leq C_5$)	*n*-BuLi, $CaZnEt_4$, $SrZnEt_4$, Na naphthalene in toluene or THF	High polymers in fairly good yields at low temperatures	184
Methyl α-ethylacrylate	*n*-BuLi in toluene and THF at various temperatures	Tacticity depends on polymerization temperature and nature of solvent; no polymer obtained > 30°C	185
D-α-Methylbenzyl methacrylate	*n*-BuLi in toluene/THF mixtures at – 78°C	Microstructure studied by NMR techniques. Relation between the tacticity and optical rotation was investigated	186
Methyl methacrylate	Na naphthalene in THF	High molecular weight; some termination due to impurities	187
	Sodium aromatic complexes in THF	High polymer	188
	n-BuLi and Na naphthalene in THF	High polymer	189
	Fluorenyllithium in toluene; 10% diethyl ether at – 78°C	Isotactic polymer; some low molecular weight polymer	190, 191
	Fluorenyllithium in toluene 0.1–15% tetrahydrofuran	Isotactic polymer at low concentrations of THF; syndiotactic polymer at high THF concentrations; at intermediate THF concentrations, stereoblock polymers formed	192
	n-BuLi in toluene	Complex kinetics	193
	n-BuLi in toluene	Complex kinetics; some low molecular weight (~ 800) polymer formed in addition to high molecular weight material; polymer possesses broad molecular weight distribution	194–196
	$LiAlH_4$–Et_2O, Grignard reagent, and *n*-BuLi in toluene	Isotactic polymer	197
	$LiAlH_4$–THF	Syndiotactic polymer	

(*table continues*)

TABLE 2.1 Double and Triple Bond Monomers (*continued*)

Monomer	Initiator system	Remarks	References
Methyl methacrylate (*continued*)	*n*-BuLi in toluene (– 30°C) *n*-diphenylhexyllithium (– 30 and – 78°C); 10% ether	Li methoxide isolated; fluorenyllithium and *n*-BuLi-initiated systems give rise to a bimodal molecular weight distribution whereas *n*-diphenylhexyllithium does not; this may be due to the large amount of alkoxide formed in the early stages of polymerization when the former initiators are used	198, 199
	n-Diphenylhexyllithium in toluene	First-order rates with respect to monomer and initiator concentration	200
	Fluorenyllithium	High polymer	201
	Na biphenyl in THF	Monodisperse polymer	202
	n-BuLi in hexane	Side reactions between initiator and monomer studied	181
	Na naphthalene in THF	Slow termination reported	203
	Na tetramer of α-methylstyrene		
	Metal amides in liquid NH_3 at – 33°C	High polymer	204
	Phenyllithium in diethylether; also MMA added to polystyryllithium in THF (– 65°C)	No polymerization; attack only on ester function; addition to polystyryllithium inclusive	205
	n-BuLi and fluorenyllithium in toluene (– 70°C); also, fluorenylsodium and Grignard reagents at various temperatures	Isotactic polymer; addition of ethers disrupts structure; mixed results with Grignards; tacticity measurements by infrared	206
	Grignard reagents; room temperature and – 40°C	Crystallizable polymer; type dependent on polymerization temperature	207
	UV and Co-60; lithium dispersion or Grignard reagents in toluene	Radiation polymer has different structure than organometallic initiated; series of crystallizable polymers prepared from isopropyl, *sec*-butyl, *tert*-butyl, cyclohexyl, and isobornyl acrylates; polymer from methyl, ethyl, and higher *n*-alkyl acrylates not crystalline; high polymer; X-ray diffraction work	208

EtMgCl–Benzalacetophenone	Crystallizable polymer with narrow MWD ($M_w/M_n \sim 1.5$)	159
Organomagnesium compounds with Mg – N bond	Stereospecific polymers	210–211
BuLi in toluene	The addition of $\geqslant 20\%$ syndiotactic PMMA to monomer gives a polymer containing 65% syndiotactic PMMA and only 2.6% isotactic PMMA	212
BuLi in toluene at $-78°C$	Isotacticity increases as monomer concentration decreases	213
Ba, Ca, or Sr metal	The microstructure of polymer varied with the dielectric constant of the reaction medium	214
ΦMgBr in toluene	Temperature dependency of the tacticity	215
t-BuONa in benzene at 20°C	Reaction carried out in the presence of *p*-benzoquinone	216
BuLi, *t*-BuLi at low temperature in THF or toluene	Branching occurs after long aging period by the reaction of the growing center of the PMMA chain with the ester group of another polymer chain	217
BuLi	Effect of ditertiary diamines on the conversion rate, MW, and microstructure of polymer	218
Dimsylsodium	Low MW	91
MeONa/MeOH at 40 and 100°C	Oligomers produced; initiation consists of an attack of CH_3O^- anion on the monomer	172
BuLi in toluene at -40 to 80°C	Kinetic study and mechanism of polymerization	220
BuLi in toluene	Reaction rate decreased by the addition of mercaptides	221
Sodium α-methylstyrene tetramer in THF at $-80°C$	Remarkable disturbance is observed in the early stage of the polymerization	222
MeONa	NMR study of the oligomers	223
BuMgBr in toluene at $-50°C$	Tacticity of the polymer affected by the absence or presence of preformed isotactic PMMA	224
Cumylcesium in THF	Estimation of the rate constant of termination by the monomer	225

(*table continues*)

TABLE 2.1 Double and Triple Bond Monomers (*continued*)

Monomer	Initiator system	Remarks	References
Methyl methacrylate (*continued*)	Fluorenyllithium	Study of the complexing effect of ether. A weakly solvating ether does not affect the isotactic character of the anionic propagation in a hydrocarbon medium, but if the donor is a crown ether, Li^+ is strongly complexed and the propagation becomes syndiotactic	226
	Li, Na, and K ketyls	Side reactions lead to formation of γ-lactones	227
	Dianion of α-methylstyrene tetramer and Na naphthalene in THF	No side reaction at $-78°C$, but at about $-65°C$, > 30 mole % of the initiator was consumed by carbonyl attack in the initiation step. At room temperature side reaction occurs in both initiation and propagation steps	228
	Fluorenylalkalis in amine media	Predominantly syndiotactic polymer	229
	NaOMe in MeOH at 70–75°C	Oligomers of 6–20 units	230
	Li diethylamide in ether, THF, or THF/benzene	Initiation reaction is slower than propagation	231
	Cumylcesium, α-methylstyrylcesium, α-methylstyrylsodium in THF at $-75°C$	No side reactions	232
	Φ_3CCaCl, fluorenylcalcium chloride, diindenylcesium, dicyclopentadienylcalcium, Φ_2Ca in 1,2-dimethoxyethane	Highly syndiotactic polymer	233
	Potassium–graphite inclusion compound in ethereal and hydrocarbon solvents	High polymer	234
	Alkali metal thiophenoxides and phenoxides in aprotic solvents	Reactivity of initiators in different polar solvents	235
	α-Lithiocarboxylic acid esters or their adducts with the alkali metal alkoxides in toluene at 20°C	α-Lithio esters are effective initiators even at 20°C in contrast with BuLi	236, 237

	Fluorenylalkali metals in ether or $(Me_2N)_3PO$	Syndiotactic polymer	238
	Cumylcesium or α-methylstyrylcesium in THF at − 75°C	Reaction proceeds without any detectable side reaction; narrow MWD	239
	t-BuOK in Me_2SO	Polymerization rapid and complete at 25°C; initiation believed to occur via the $MeSOCH_2^-$ ion with the monomer	240
	Li biphenyl in dimethoxyethane	Syndiotactic polymer; bimodal MWD	241
	Na 1,3-diphenyl-1,2,3-propanetrione in 1,2-dimethoxyethane	Does not induce anionic polymerization; effective purifying agent in the removal of H_2O, O_2, etc.	242
	Benzyl sodium in 1,2-dimethoxyethane or THF	Slow polymerization	242
	Na naphthalene, Na anthracene, Na triphenylmethane, Na or Li benzophenone, or the Michler ketone Na salt	Rapid polymerization	242
	9-Fluorenyllithium in 5–10% THF in toluene (− 60°C)	Stereosequence distribution	243
	Li biphenyl in aromatic hydrocarbons	97% syndiotactic polymers	244
	Na or K in Me_3N	Kinetics of polymerization	245
Oligoester acrylate	Na naphthalene	Study of electron exchange between catalyst and monomer	246
Phenyl acrylate	BuLi in toluene at − 50°C	Isotactic polymer	247
Phenyl methacrylate	*n*-BuLi in toluene and THF	Isotactic polymer prepared in toluene at − 50°C	248
β-Styryl acrylate	Optically pure 2-methylbutyllithium	Optically active polymer	249
2,2,6,6-Tetramethyl-4-piperidinol-1-oxyl methacrylate	ΦMgBr in toluene or chloroform at 0°C	MW ~ 1000; paramagnetic polymer	250
p-Tolyl methacrylate	BuLi, Na naphthalene, $LiAlH_4$ in toluene and THF	Essentially isotactic polymer in toluene, but syndiotactic in THF	251
Trityl methacrylate	*n*-BuLi, (ΦMgBr)	A highly isotactic polymer in toluene and THF; no influence by the solvent	166 252

(*table continues*)

TABLE 2.1 Double and Triple Bond Monomers (*continued*)

Monomer	Initiator system	Remarks	References
1,3-Dienes			
Butadiene	2-Phenylisopropylsodium in ether	High molecular weight	253
	Sodium	Heterogeneous initiation; high polymer obtained; anionic nature of reaction demonstrated	254
	Sodium initiation	Heterogeneous initiation; high polymer	255–263
	Sodium in toluene	Ion–radical nature of initiation indicated; propagation reaction anionic; low molecular weight caused by chain transfer to toluene	264
	Tritylsodium	Very slow initiation	265
	EtLi, Na, and K in hydrocarbons, tetrahydrofuran, and triethylamine	Microstructure analysis	266
	n-BuLi in hydrocarbons	Kinetics of initiation and propagation	55, 267–269
	1,4-Dilithiobutane in benzene/Et_2O	Polybutadiene glycol was prepared by Et_2O termination	270
	Dixanthenylbarium in $CH_3OCH_2CH_2OCH_3$	Low MW ~ 1850; high 1,4 configuration (73%)	271
	BuLi in *n*-hexane	Polymerization in a system having a low solvent–monomer ratio (2–4:1)	272
	BuLi in aliphatic solvents at 25°C	The first monomer units are incorporated largely by 1,2-addition. This extent decreases with degree of polymerization	273
	BuLi in aliphatic solvent at 25°C	The effect of some Li salts on the microstructures of low MW polydienes was determined	274
	Mixtures of BuLi and *t*-BuOLi, *s*-BuOLi, *iso*-BuOLi, EtOLi in toluene	Polymers with low viscosity and high 1,2-addition product	275
	n-BuLi in benzene	Kinetic studies	209
	1,4-Dilithio-1,1,4,4-tetraphenylbutane at 80°C in C_6H_6–Et_2O	Bifunctional polymer with > 80% 1,4-microstructure and MW > 6000	276
	s-BuLi in isooctane at − 40° and 30°C	Spectral interpretation of the nature of the active center in the living polymer	277

s-BuLi in isooctane at − 78°C	Spectral study; addition of ether reduces the association of polymer	278
BuLi, *s*-BuLi, *t*-BuLi in hydrocarbons	The nature of the deactivator, solvent, and initiator as well as the monomer/initiator ratio affected the microstructure of oligomer (DP 1–3 and 9)	279
s-BuLi in heptane, cyclohexane, and toluene at 20–60°C	Polymerization kinetics; activation energy of propagation and rate constants	280
Organobarium in THF	Conductometric study of the polymerization; copolymerization with styrene reported	281
1,1-Diphenylethylene dimer barium salt in ether–hydrocarbon	The influence of solvent on microstructure of polymer	282
s-BuLi in polar and nonpolar media	High ratio of initiator to monomer (2:1 and 1:1); NMR study of the oligomer	283
s-BuLi	Oligomers formed were hydrogenated and characterized by gas chromatography	284
s-BuLi in hexane	Mode of monomer addition with respect to the polarity of the medium	285
t-BuLi in benzene	NMR study of the structure of oligomeric polybutadienyllithium and polybutadiene	286
Alkyllithium	The nature of the propagating chain end	287
t-BuLi in hydrocarbon solvent	NMR study of the 5-BuLi/1,3-butadiene (1:1) product	288
BuLi in heptane	Allyl bromide or 1,4-dibromo-2-butene was used to control the MWD	289
Na in THF, kerosene, and *n*-hexane at − 46 to − 74°C	Living polybutadiene	290
Li, Na, K, Mg, Ca, and Ba aromatic complex in $(Me_2N)_3PO$	Polymerization mechanism involving an allyl ion proposed	291
Na, Na naphthalene, or Na biphenyl complex	The MW of the polymer decreased when a 1-alkene was used as the solvent instead of benzene (due to chain transfer reaction)	292

(*table continues*)

TABLE 2.1 Double and Triple Bond Monomers (*continued*)

Monomer	Initiator system	Remarks	References
Butadiene (*continued*)	BuLi in ΦCl at 50°C	A remarkably rapid reaction rate is obtained in a medium containing > 30 wt % chlorinated or fluorinated aromatic diluent	293
	n-BuLi in cyclohexane	Influence of temperature on microstructure	294
	BuLi in hexane	The presence of polar modifiers affect the microstructure	295
	BuLi	Thermal stability of the living polymer	296
	EtLi, *t*-BuONa, *t*-BuOK	IR study of the nature of active centers	297
	BuLi in cyclohexane	The presence of alkali alkoxides drastically increases the rate of polymerization and vinyl unsaturation in polymer	298
	Li, Na, K, or the corresponding oligomeric polystyryl salt in THF	UV study of the isomerization of polybutadienyl salt in the presence or absence of the corresponding tetraphenylboride	299
		The ion pair propagation rate determined	300
	BuLi/1,2-dimethoxyethane in hexane at 58°C	Polymers with vinyl content 20–50%	301
	Dibenzylmagnesium/tetrahydrofurfuryl alcohol in THF at − 70°C	Effective initiator; no polymerization induced by the magnesium alkyl alone	302
	Li dispersion in cyclohexane at 70°C	Polymers contain high 1,4-addition product	303
	Alkali metals, alkali metal *t*-butoxides in hexane	Alkali metal alkoxides are used as modifiers; the microstructure of polymers varies with the type of modifier and polymerization temperature	304
	Alkylsodium/crown ethers; Li in hydrocarbon solvents	Polymers with high vinyl content (80%) in high conversion	305
	Barium salt of 1,1-diphenylethylene dimer in ethereal and hydrocarbon solvents	The polymer is more stereospecific than with initiators based on other metals of groups I and II	306
	RLi/dialkylzinc in cyclohexane at 70°C	Dialkylzinc complexed with the initiator and effectively increased the initiation rate	307

	Organolithium substituted with hydroxyl-carrying mixed acetals	Acetal-terminated polymers	308
	Li acetylide	Broad MW with high (85–95%) 1,4-addition product	309
	1,2-Dilithio-1,2-diphenylethane in *n*-pentane	Polymer with 44.8% cis structure	310
	BuLi in hexane at 50°C	The living polymer was treated with $TiCl_4$–$Et_3Al_2Cl_3$ to increase MW	311
	Na dispersion in 1,4-dioxane/heptane mixture	The addition of $(Me_2N)_3PO$ accelerates the polymerization reaction; polymer highly 1,2-addition product	312
	Dixanthenylbarium in 1,2-dimethoxyethane at 25°C	Polymer with high 1,4-product (73%)	313
	n-BuLi in hexane and H_4-furan	Kinetic studies	49, 51, 209
1,1-Bis(trifluoromethyl)-1,3-butadiene	BuLi in THF at − 78°C	Crystalline polymer	314
Chloroprene	BuLi in hydrocarbon solvent	Low degree of conversion (~ 9%) possibly due to termination	315
	Bu_2Mg/BuMgI in heptane	Kinetics study	316
	BuLi · LiI–$MgBu_2$, LiBu · LiI–LiBu LiBu · 2LiBu in heptane	80–90% conversion with polymer having trans-1,4 configuration	317
	BuLi or Bu_3Mg_2I	Kinetics study	318
1-Cyanobutadiene, *trans*	BuLi in toluene or THF	Polymer of predominantly trans-1,4 configuration; copolymerization with methyl methacrylate and methyl pentadienoate reported	319
	BuLi or *n*-octylsodium in hydrocarbon solvents at − 78°C	Polymers with substantially trans-1,4 structure obtained	320
1,3-Cyclohexadiene	BuLi in toluene ⩽ 100°C	Poly(1,3-cyclohexadiene)	321
2,3-Dimethyl-1,3-butadiene	BuLi in heptane or Et_2O	Polymerization rate increased with increasing pressure	322
	BuLi in THF	1,2 units decreased from 84% to 48% with increasing polymerization temperature from − 78 to − 30°C	323
	BuLi in THF	NMR study of polymer microstructure	324
	t-BuLi, *s*-BuLi in hydrocarbon solvent	Reaction orders in different solvents	325

(*table continues*)

TABLE 2.1 Double and Triple Bond Monomers (*continued*)

Monomer	Initiator system	Remarks	References
2,3-Dimethyl-1,3-butadiene (*continued*)	*n*-BuLi	The influence of various parameters (temperature, solvent type, initiator concentration, pressure) upon the microstructure of polymer	326
	BuLi in hexane, Et_3N, bulk, or THF	Kinetics of polymerization and microstructure	327
	C_2D_5–benzene-d_6	NMR study of active chain ends	9, 327a
2,3-Dichloro-1,3-butadiene	*iso*-BuLi in *n*-heptane	A tough solid, insoluble in benzene, predominantly 1,4 configuration	328
2,4-Hexadienoic acid (sorbic acid) alkyl esters	Optically pure 2-methylbutyllithium	Optically active polymer	249
1,3,4-Hexatriene	Soluble VCl_3–$Al(i\text{-}C_4H_9)_3$	High molecular weight, linear polymer; 1,2-, 1,4- and 1,6-enchainment found; the trans–trans, the cis–trans, the conjugated and nonconjugated vinyl, and the trans groups found in 1,4- and 1,6-polyhexatriene	329
	BuLi in benzene at 0–30°C	Unsaturated reactive linear polymer of MW 1,000–7,000 prepared	330
Isoprene	Na metals	—	331, 332
	Li, Na, Cs, Rb metals,	—	333
	Li metal in hydrocarbons	High cis-1,4 content	334–336
	s-BuLi–diisopropenylbenzene in hydrocarbon solvent	α,ω-Dilithiopolyisoprene of high cis-1,4 content was prepared	337
	BuLi in anisole	The cis-1,4 content of the polymer decreased with increasing pressure, apparently reflecting the augmented ionic character of the polymerization	338
	n-BuLi in nonpolar and polar solvents	The microstructure of polymer prepared in hydrocarbon solvents is very sensitive to pressure, whereas in oxygenated solvents the structure remains unchanged	339
	Butylisoprenyllithium in *n*-heptane and *n*-heptane–diethyl ether mixtures at 20°C	Effect of ether on the rate of polymerization	340

Tertiary N-substituted allylamines containing Li	Polymers with functional groups prepared	341
s-BuLi in *n*-hexane	The initiation step was studied. The effect of *t*-BuOLi on the rate was also investigated	342
s-BuLi in hydrocarbon solvent at 30–45°C	Successive addition of initiator to increase the workability of the polymer	343
s-BuLi in cyclohexane	Different kinds of association exist during polymerization	344
n-BuLi in heptane	Mechanistic study and MWD	345
EtLi, *s*-BuLi in benzene, *n*-hexane and cyclohexane	The states of association were studied	346
EtLi, *s*-BuLi in benzene	The cross-association of ethyllithium and polyisoprenyllithium	347
s-BuLi in cyclohexane	States of association	348
s-BuLi or *t*-BuLi in benzene at 10°C	The nature of active species in anionic polymerization studied by NMR spectroscopy	349
BuLi in benzene	Dyad distribution of cis-1,4 and trans-1,4 units by ^{13}C NMR spectroscopy	350
RLi/TMEDA	A significant increase of polymerization rate and of the 3,4 content when RLi/TMEDA ≈ 1	351
Polystyryllithium, Na naphthalene, Na α-methylstyrene tetramer, or polystyrylsodium in THF at − 40 to 0°C	Stability of the carbanions depends on the type of counterion; isomerization occurs due to instability of carbanion	352
Li diethylamide in hydrocarbon media and polar solvents	Effect of additives on the rate of polymerization	353
Organoalkali metals in dioxane	Microstructure of polymer; transition state in propagation step	354
Li, Na, K, Cs catalysts in various solvents	Dependence between the stereospecificity of the anionic propagation of isoprene and the nature of the active centers. Influence of polymerization temperature, nature of solvent on the propagation mechanism	355
Graphite–alkali metals	Microstructure depended on the nature of the active centers	356

(*table continues*)

TABLE 2.1 Double and Triple Bond Monomers (*continued*)

Monomer	Initiator system	Remarks	References
Isoprene (*continued*)	(1,1-Diphenylhexyl)lithium in hexane or benzene	Microstructure of polymer formed is independent of the MW or the solvent, but dependent on temperature	357
	s-BuLi or *t*-BuLi in benzene	NMR study of the oligomeric poly(isoprenyllithium) (DP 1.2–3)	358
	n-, *s*-, or *t*-BuLi in cyclohexane	Polymerization in the presence or absence of the corresponding lithium butoxides	359
	1,2-Dilithio-1,2-diphenylethane	Polymer with 72% cis structure	310
	MC_n where $n = 16, 40$ when M = Li, $n = 24, 32$ when M = K in THF and cyclohexane	High vinyl contents	360
	EtLi in toluene	Polymerization rate and MW were affected by the addition of anthracene	39
	s-BuLi; polybutadienyllithium in benzene-d_6	NMR study of propagating chain end	9, 327a
	n-BuLi in hexane, benzene, cyclohexane and H_4-furan	Kinetic studies	49, 51, 209
1,3-Pentadienes and 2,6-hexadiene	*s*-BuLi in benzene-d_6	NMR study of propagating chain end	9, 360a
1-Methylcarboxyisoprene	BuLi in toluene at − 60 to − 70°C	80% yield in ⩽ 60 min; crystalline polymer with trans-1,4 configuration (IR)	361
1,3-Pentadiene	BuLi in $(Me_2N)_3PO$ or THF	Solvent effect on the microstructures of the polymers	362
	BuLi in heptane at 80°C	The structure of the polymer backbone was determined by ^{13}C NMR spectroscopy	363
trans-2-Methyl-1,3-pentadiene	BuLi	Oligomers showing regular 4,1-coupling	364
4-Methyl-1,3-pentadiene	Na or K	The nature of solvent and reaction temperature did not affect the structure	365
	BuLi in hexane	Addition of diisopropyl ether increased the content of isobutenyl side group	365
	BuLi in benzene or heptane	Mainly 1,4-addition products	366
	Li, Na, K, BuLi	Effect of ionic character of C – M bond on microstructure	367

1,3,5,7-Octatetraene	BuLi in benzene at 0–30°C	Unsaturated reactive linear polymer of MW 1,000–7,000 prepared	330
2-Phenyl-1,3-butadiene	*t*-BuLi in hydrocarbon media	Oligomer obtained	368
	BuLi in cyclohexane/anisole or THF	Influence of polar modifier such as $(Me_2N)_3PO$, on microstructure of polymers	369
	Na in dioxane at 62°C	Also used 1-phenylbutadiene	370
	Cumylpotassium, Li naphthalene, Na naphthalene in THF at 0°C	Cis 1,4-addition; 100% conversion	371
trans-(*R*) (−)-5-Phenyl-1,3-hexadiene	BuLi in heptane	Only 1,4-structure present by IR and NMR; optically active polymer	372
trans-(R)(−)-6-Phenyl-1,3-heptadiene			
Ethylene and Vinyls			
Ethylene	*n*-BuLi in hydrocarbons, 420–970 atm	Low molecular weight	373
	n-BuLi complexed with tetramethylethylenediamine	High molecular weight polymer	374, 375
	n-BuLi in ether, 100–500 atm	Low molecular weight	376
	t-BuLi and isopropyllithium in ether, 1 atm	Initiation but no polymerization	377
	n-BuLi spartine complex	Low molecular weight produced ($\overline{M}_n = 1300–1400$) in octane	378
	BuLi–TMEDA in hexane	Formation of oligomers	379
		Kinetics of polymerization	
	BuLi–TMEDA in *n*-hexane	The free monomeric BuLi, unchelated to TMEDA, is the active initiating species	380
	BuLi–TMEDA	Other chelating agents were considered and TMEDA was the most effective; MW up to 10^5	375
	BuLi–TMEDA	Oligomerization at low pressure and temperature	381
	BuLi–TMEDA in benzene	Telomerization to a waxy material	378

(*table continues*)

TABLE 2.1 Double and Triple Bond Monomers (*continued*)

Monomer	Initiator system	Remarks	References
Ethylene (*continued*)	*t*-BuLi/TMEDA at 0°C	Oligomerization at low pressure; the rate of consumption of ethylene depends on the pressure, *t*-BuLi concentration, and TMEDA concentration; functionalized oligomers obtained by adding acetone at − 78°C	382, 383
	t-BuLi–TMEDA	Living polymer functionalized by addition of excess acetone at − 78°C	384
Acenaphthylene	*n*-BuLi in THF	Low molecular weight	385, 386
Acrylamide	RLi in toluene	Crystalline, tactic polymer; high conversion	387
	t-BuONa in *o*-dichlorobenzene at 120°C	High yields (95%) of poly(α-alanine)	388
	BuLi, BuLi–$TiCl_4$ at − 70°C	Polyacrylamide or a polymer containing poly(β-alanine) unit were obtained	389
	s-BuONa, $(s\text{-BuO})_2Ca$ in aromatic solvents, dioxane, pyridine, $HCONMe_2$	Effect of solvents on chain structure	390
	s-BuONa in polar solvents, such as pyridine, $HCONMe_2$, ΦNO_2	Effect of solvents on chain structure	391
	Alkali metal methoxides in methanol and in aprotic solvents	Oligomerization reaction studied	392
	t-BuONa in *N*-methyl-2-pyrrolidone at 100°C	Hydrogen transfer polymerization; branching in polymer studied	393
	t-BuONa in aprotic solvents	Chain propagation mechanism	394
	t-BuONa in xylene at 50–60°C	High yield of poly(β-alanine)	395
	t-BuOK in sulfolane at 100°C	Hydrogen transfer polymerization; mechanism	396
Acrylamides (methyl substituted) (acrylamide, crotonamide, methacrylamide, *N*-methylacrylamide)	*n*-BuLi in pyridine at 110°C	Effect of methyl substituent on rate of polymerization and MW	397
Acrylonitrile	RLi, RNa, ROLi, RONa, Na naphthalene	Generally no termination; polymerization in nonsolvent limits molecular weight; high molecular weight polymer obtained in toluene or THF/DMF mixtures; some branched material formed	398–406

Acrylonitrile (*continued*)	Bu_3Mg_2I in toluene	The activating effect of the Lewis bases $HCONMe_2$, Me_2SO, and $(Me_2N)_3PO$ was studied	407
	BuMgCl in toluene at − 75°C	Kinetics study; termination free	408
	Bu_3Mg_2I–DMF at 75°C	Catalytic effect of DMF	
	Bu_2Mg, BuMgCl	Almost monodisperse polymer obtained	409
	Bu_3Mg_2I	Broad MWD due to the coexistence of different growing centers of polymerization	
	RSM where M = Li, Na, or K	Polymerization in both polar and nonpolar solvents	410
	BuLi in toluene at −70°C	~50% of the initiator is used in the formation of acrylonitrile oligomers	411
	Metal alkyls in hydrocarbon media	Oligomerization accounts for the low initiation efficiency	412
	ROC(=S)−S−M where M = Li, Na, K, and Zn in DMF at 0°C	The initiation ability of the xanthates increases with increasing basicity of the alcohol of the alkoxide group and with decreasing electronegativity of the metal ion	413
	Et_3P in DMF	Polymerization via macrozwitterions	414
	Φ_3P in DMF at 30°C	Initiation by $CH_2{=}C^-(CN)$ species	415
	BuLi, Li butyldipropylcarbinolate in toluene or DMF at low temperature	Low effectiveness of the catalyst due to side reaction	416
	BuLi in hexane and THF	*n*-Butane formation and nitrile addition are important under these conditions	417
	Dimsylsodium	MW < 20,000	219
	BuLi in toluene at −75°C	Low efficiency of initiator due to oligomerization	418
	R_3P in $HCONMe_2$ at 25°C	The effect of LiCl on the polymerization was studied	419
	Bu_3P	Unsaturated oligomers obtained	420
	Et_3P	Kinetics and mechanism of polymerization	421
	Diphenylsulfoxide potassium (1:1 or 1:2) in THF at −78°C	Effective initiator	422
	Polybutadiene metal derivatives	Block copolymer prepared	423

(*table continues*)

TABLE 2.1 Double and Triple Bond Monomers (*continued*)

Monomer	Initiator system	Remarks	References
n-Alkenamide $CH_2{=}CH(CH_2)_xC(=O)NH_2$, $x = 2, 3, 4$	*t*-BuOK in DMSO, DMR, dioxane, pyridine	3-Butenamide and 4-pentenamide give polypeptides by isomerization polymerization; 5-hexenamide does not polymerize	424
α-Alkenylphosphonic acid derivatives	Et_3Mg_2Br in bulk polymerization	Yield ranging 40–80%; transparent polymers which readily absorb H_2O or MeOH	425
α-Alkyl acrylonitriles	MeMgBr in THF or toluene	Yellow polymer; no polymer formed when R ≥ C	426
t-Butyl crotonate	RLi in toluene or ether	Low molecular weight and amorphous	427
	2-Methylbutyllithium	Living polymerization, monodisperse, soluble polymer	428
Crotonaldehyde	R_3P in various solvents	Low molecular weight	429
Crotonic esters	$LiAlH_4$ in ether at −70°C	Oily low MW polymers for R = Me, *iso*-Pr, *iso*-Bu and high polymer for R = *s*-Bu	430
	Alkali metal naphthalene, ΦMgBr, $LiAlH_4$	No polymerization for *cis*-*s*-Bu crotonate, but high polymer for trans isomer	431
Crotononitrile	Alkali metal naphthalenes, ΦMgBr, $LiAlH_4$	The trans isomer is more reactive than the cis isomer	431
Cyanoacetylene	Amines, *n*-BuLi, RO^-, CNS^-, and CN^- in DMF at 0°C	Low molecular weight polymer ($\overline{M}_n = 500$–600); polymerization through carbon triple bond	432
	NaCN in Me_2SO_4	Brown to black amorphous polymer, soluble in polar organic solvent $(CH{=}C(C{\equiv}N))_n$	433
Diacetone acrylamide	Na naphthalene in THF or *N*-phenyl-2-naphthylamine/BuLi mixture or *N*-phenyl-2-naphthylamine/ΦMgBr mixture	Polymer soluble in acetone and precipitated in H_2O as a viscous oil	434
2,4-Dicyano-3-alkyl-1,4-pentadiene	EtLi, MeMgBr, Na ketyl in toluene or THF	Low MW, yellow polymer formed	426

Dicyanoacetylene	*n*-BuLi, Li naphthalene, and Li benzophenone in THF at $-75°$ and $0°C$	Low molecular weight polymer ($\overline{M}_n = 400–500$); low molecular weight is caused by the gradual degeneration of the active center as a result of resonance stabilization in the conjugated system	435
Diethyl methylenemalonate	Φ_3P	Polymerization through a zwitterion mechanism	436
1,2-Dihydronaphthalene	Na naphthalene in diethyl ether	Powdery polymer	437, 438
	Na biphenyl in THF, *n*-BuLi in THF or benzene	Low molecular weight, insoluble polymer	439
	Na polycyclic compounds	Living polymer; block copolymer with styrene reported	440
	BuLi in aromatic hydrocarbons and ether		441
p-Diisopropenylbenzene	BuLi in THF or benzene at 20–40°C	IR analysis of the oligomer structure indicated only one of the isopropenyl groups participated in polymerization which occurred through the vinylidene group	442
Dimethyl itaconate	Alcoholic alkali metal and magnesium methoxides in MeOH, DMF, and DMSO	Yield of oligomers higher in aprotic solvents; oligomers contain no MeO^- end group	443
Diphenylacetylene, diphenylbutadiyne	K, Na dispersion, Li naphthalene, Na naphthalene, or BuLi at -20 to $+65°C$	Low MW, soluble, colored, paramagnetic polymers	444
1,1-Diphenylethylene and *trans*-stilbene	Dibenzylmagnesium in $(Me_2N)_3PO$	Oligomer	445
p-Di(1-phenylvinyl)benzene	Na, K, Na naphthalene	Oligomers up to the decamer are formed	446
Diphenylvinylphosphine oxide	*n*-BuLi, Na naphthalene, in toluene, THF, and pyridine	Low molecular weight	447
N-ethyl-2-vinylcarbazole (and *N*-ethyl-3-vinylcarbazole)	*n*-BuLi, cumylpotassium, Na biphenyl, α-methylstyrene tetramer dianion, in benzene, THF	2-Vinyl monomer exhibits all the features of a standard living polymer; 3-vinyl is less stable, i.e., only at $T \leqslant -40°C$	448
Ferrocenylacetylene	Na at $>130°C$ in bulk polymerization	10% yield; MW ~ 1700	449
2-Isopropenyl-4,5-dimethyloxazole 2-Isopropenylthiazole 2-Isopropenylpyridine	Na naphthalene or Na-α-methylstyrene tetramer in THF		450

(*table continues*)

TABLE 2.1 Double and Triple Bond Monomers (*continued*)

Monomer	Initiator system	Remarks	References
2-Isopropenylthiazole	*n*-BuLi, bulk; $MgBr_2$, toluene at 0° and 25°C	Soluble polymers	451
4-Vinyl- or 4-isopropenylthiazole	*n*-BuLi, bulk or hexane at 25°C	Traces of polymer	451
Maleamide Mesaconamide	BuLi or *t*-BuONa	Oligomeric polypeptides	452
Maleimides (N-substituted)	*n*-BuLi in toluene and tetrahydrofuran	Low temperature (−78°C); threo diisotactic configuration	453
Methacrylamide	*s*-BuONa in polar solvents	Transition polymerization occurred	391
Methacrylonitrile	Lithium in NH_3	No termination	454
	Sodium in NH_3	High molecular weight	455
	n-BuLi in toluene, −78°C	Mixture of isotactic (32%), syndiotactic (13%), and heterotactic (55%) blocks	456
	Et_3P in THF	The polymerization rate is slowed down by polymethacrylonitrile, water or $Me\Phi_3P^+I^-$; reaction proceeds by a zwitterionic mechanism	457
	BuLi in hexane and THF	*n*-Butane formation and nitrile addition are important under these conditions	417
	s-BuLi in benzene	The nature of the active centers	458
	K and Na α-methylstyrene oligomers, *s*-BuLi in THF at −70°C	The active centers are able to attach to the cyano groups of some aromatic nitriles and to the carbonyl of benzophenone (added)	459
	Et_3P in DMF	Termination reaction by water	460
	Et_3P	Anionic zwitterion-type polymerization	461
N-Methyl-α-methylbenzene-acrylamide	*n*-Butyllithium in toluene at 22° and −50°C	Optically active polymer	462
2-Methyleneglutaronitrile	Li naphthalene (−35°C) or diphenylmethanelithium (−70°C) in DMF	Nearly white polymer (yield >75%) with broad MWD and high degree of branching	463
	NaCN in DMF at −42 to −32°C	Colorless polymer at low yield	464
Nitroethylene	RNH	—	465

2-Nitrobutene	*n*-Butyllithium in DMSO; lithium in DMF and DMSO; $LiAlH_4$ in toluene; NaH in DMSO; Na naphthalene in THF; NaCN in DMF and CH_2Cl_2	High conversions and low molecular weight polymer at 25°C; low temperatures (~ −78°C) may lead to the formation of high molecular weight material	466
1-Nitropropene	Potassium bicarbonate	High polymer	467
	BuLi in toluene, THF, or DMF	Soluble amorphous polymer with MW ~ 38,000	468
	MeONa	Partly insoluble, with some degree of ordering	468
Phenylacetylene	*n*-BuLi in heptane/$(Me_2N)_3PO$	Polymerization kinetics. Broad MWD and presence of C≡C in polymer due to chain transfer reaction	469
	BuLi, Li phenylacetylenide in $(Me_2N)_3PO$	Intra- and intermolecular chain deactivations occur	470
	BuLi in hexametapol	Polymer with MW 2500	471
Propiolates (carboxyacetylenes): Methyl propiolate HC≡C−COOCH; Methyl acetylenedicarboxylate; Methyl phenylpropiolate; Ethyl tetrolate; Acetic propiolic anhydride	Halides, KSCN or *Φ*SNa in DMF	The reactivity and mechanism investigated	472
N-Propyl-*N*-α-methylbenzyl-acrylamide			
Trimethylsilylvinylacetylene	BuLi	Polymerizes through vinyl group to white, soluble polymer (high yield)	473
	Li-naphthalene, Na naphthalene, in THF at −78°C		
Vinyl acetate; vinyl monochloroacetate, -propionate, and -benzoate	Lithium metal in bulk, benzene, and THF	Polymer of moderate molecular weight apparently obtained	474
Vinylalkyl(aryl)silanes	RLi in heptane or toluene −60° and +75°C	Thermodynamics of polymerization; high MW product	475
		High MW product (high yield)	476
		Soluble polymer of high T_g	477

(*table continues*)

TABLE 2.1 Double and Triple Bond Monomers (*continued*)

Monomer	Initiator system	Remarks	References
1-Vinylanthracene	Na α-methylstyrene tetramer in THF (−78°C)	Soluble polymer, MW ~ 10^5	478
2-Vinylanthracene (and 2-propenyl-2-anthracene)	Na α-methylstyrene tetramer (THF), −78°C	High polymer of broad MWD	478
	BuLi (benzene)	Oligomer	
	Na naphthalene, Na biphenyl		
9-Vinylanthracene	Na biphenyl, Na naphthalene, BuLi, Na α-methylstyrene tetramer, living polystyrene in THF or benzene at −78 to +25°C	Oligomers	479
Vinyl aryl monomers $CH_2{=}CH(R)C_6H_4R'$ (11 monomers)	Na naphthalene, Na benzalaniline, *n*-BuLi in THF	Some give crosslinked polymers, but most give linear polymers. (Cationic or radical give crosslinked polymers in all cases)	480
p-Vinylbenzamide	K *t*-butoxide in DMF and DMS at ~ 110°C	Repeating unit is 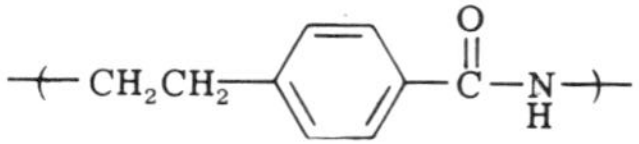 45% yield; ortho and meta isomers give no polymer and low yield, respectively	481
Vinyl chloride	RLi in hydrocarbons or THF	Low molecular weight; no polymer in THF	482
	BuLi, BuLi–BuI in heptane at 0–40°C	Mechanistic study	483
	t-BuMgCl	The monomer and catalyst were used up at equal rates, and polymerization ceased when all the initiator was used up	484
	Alkyllithium at −20 to +20°C	Effect of temperature, initiator concentration, and monomer concentration on the conversion and the properties of the resulting polymer; MW of polymer as high as 140,000	485
	t-BuMgCl in THF and other solvents	Simultaneous participation of anionic and radical mechanisms	486

	t-BuLi at 0°C	Low yield (38%) with M_n = 55,000	487
Vinyl ethers	Grignard reagents	High molecular weight polymer; O_2 and CO_2 serve as cocatalysts	488, 489
Vinyl isocyanate	NaCN in DMF	At −55°C, polymerization through the −NCO group; polymerization through the vinyl group at room temperature	490–492
Vinyl ketones	RLi, ROLi, or biphenyllithium	Crystalline in toluene but not in THF	493, 494
	$MgBr_2$, Et_2Zn	Highly crystalline polymers	495
	Et_2Zn in hydrocarbon solvents and ethers	Mechanism elucidation by UV and cryoscopic methods	496
	Alkylzinc	Polymerization mechanism	497
	Ethylzinc 1,3-diphenyl-1-pentene-1-olate in toluene at 0°C	The polymerization system exhibited living characteristics	498
	n-BuLi in benzene at 0°C	A bimodal MWD of low M_n; the structure of the oligomer indicates essentially normal vinyl polymerization, but there is evidence for a small proportion of conjugate addition	499
1-Vinylpyrene	Cumylpotassium, K α-methylstyrene, THF (−60°C)	Living system and copolymerization with other monomer to give block (AB and ABS) copolymers	500
Vinylpyridine	MgR_2, RMgX, R_2Be, or RLi	Isotactic polymer	500
1-Vinylpyridine	BuLi in benzene or cyclohexane cumylpotassium, α-methylstyrene tetramer dianion, naphthalene anion-radical in THF	Living polymer; also block copolymers with ethylene oxide, styrene, and isoprene	501
2-Vinylpyridine	Metal amides of Li, Na, Be, Al; organometallic compounds of Li, Na, K, Be, Zn; Grignard reagents	Isotactic polymers, acetone soluble, and crystallizable	502, 503
	Na naphthalene in THF at 25°C	High MW polymer	504
	Diphenylmethylsodium	Incomplete conversion, complicated by the high reactivity of the pyridine ring; dimerization	505

(table continues)

TABLE 2.1 Double and Triple Bond Monomers (*continued*)

Monomer	Initiator system	Remarks	References
2-Vinylpyridine (*continued*)	Cumylpotassium, Na, Cs, Na and K biphenyl; Na α-methylstyrene tetramer in THF	UV study of the initiation of the living anionic polymerization. Living polymer; conductivity measurements	506
	Na biphenyl; Na naphthalene in THF	Conductivity measurement indicates possible interaction between the Na counterion and the nitrogen atom in the pyridine ring	507
	Na polycyclic hydrocarbon	Living polymers and block copolymers with styrene	440
	Na naphthalene, Na biphenyl, and cumylpotassium in THF	Good isomolecularity, $M_n \simeq M_s$	508
	2-Ethylpyridyllithium in THF at −78°C	Oligomerization	509
	Dialkyl- and diarylmagnesium in THF at 20°C	Spectrophotometric observations indicate the existence of stable active species with different structures	510
	Na α-methylstyrene tetramer, Cs α-methylstyrene in THF	Living polymer	511
	Cumylcesium, Na α-methylstyrene tetramer in THF	Equilibrium of active centers (contact ion pairs, free ions, positive triple ion)	512
	Cumylcesium in THF	Addition of 1,1-diphenylethylene to the active species of 2-vinylpyridine; kinetic study	513
3-Vinylpyridine	Na naphthalene in THF	Living polymers	514
4-Vinylpyridine	Na naphthalene in THF at 25°*C*	High MW polymer	504
	Dimsylsodium	Low MW (≤20,000)	219
	BuLi in toluene or heptane	High MW; living polymer	515
Vinyl sulfones			
Methyl vinyl sulfone	*n*-BuLi in THF, toluene or bulk	Low yield ≤40%, polymer with MW 8000–20,000 Soluble in DMF, Me_2SO, CF_3COOH	516
	K in bis-2-ethoxylethyl ether, or Li in THF, or hexane	1,4-Bis(methylsulfonyl)butane was obtained	517

Methyl vinyl sulfone, phenyl vinyl sulfone	*n*-BuLi, $ZnEt_2$, $LiN(CH_3)_2$, $NaNH_2$, *n*-BuLi–$ZnEt_2$ in pyridine, sulfolane, toluene, or benzene	Polymers of MW ~ 10,000	518
Alkyl vinyl sulfone	Organometallic catalyst at −10 to −80°C		519
Vinyltrialkyltin	BuLi	The catalyst lost its reactivity during the process due to complex formation with the vinyltrialkyltin macroanions; low MW polymer	520
	BuLi	Deactivated centers in the polymer were reactivated by ethers; block copolymerization with styrene reported	521
Triethylvinyltin, triethylvinylgermanium	*n*-BuLi in hydrocarbons	Low MW (M_n = 5,000–9,000); powdery polymer; copolymerization with styrene reported	522
Vinylverdazyl	*n*-BuLi or Na naphthalene	Dark green polymer (MW ~ 2800); copolymer with styrene and MMA reported	523
Vinylidene chloride	RLi	High molecular weight	524
Vinylidene cyanide	ROH, RNH_2, H_2O	–	525

TABLE 2.2 Monomers Based on Carbon–Heteroatom Multiple Bonds (Heterounsaturates)

Monomer	Initiator system	Remarks	References
Acetaldazine $CH(CH_3)=N-N=CH(CH_3)$	BuLi in aprotic solvents at 0°C	Formation of oligomers with MW 800–1200 due to proton transfer from monomer to propagating anion	526
	NaOMe	No polymer obtained	526
	Grignard reagent, metal ketyl	Acetaldazine polymerized more readily than styrene or butadiene	526
	BuLi in heptane	A viscous liquid polymer with a MW of 400–1000	527
N-acylethyleneimine (1-acetylaziridine)	Et_3N at −20 to 200°C	Polymer formed dissolves in alcohols and H_2O	528
Acetaldehydes (halo substituted)			
Dichlorofluoroacetaldehyde	Φ_3P, *t*-BuOK at 30°C	Polymer nature depends on polymerization conditions and type of initiators	529
Chlorodifluoroacetaldehyde	Φ_3P, *t*-BuOK	Polymer nature depends on polymerization conditions and type of initiator	530
Trichloroacetaldehyde (chloral)	Bu_3N, Et_2Zn, 2,6-dimethoxyphenyllithium/Et_2Zn in hydrocarbon solvent	High polymer (DP 300–600)	531
Trifluoroacetaldehyde (fluoral)	BuLi at −78°C (bulk polymerization)	White elastomer	532
	CsF/tetraglyme in ether at −78°C	High MW, amorphous polymer	533
Acrolein	$CH_2=CHCH_2ONa$ at −60 to −20°C	Termination of active center at −20°C	534
	R_3P, Et_3N, MBH_4, MCN (at −78 to +20°C), MCNS, Na naphthalene, and Li naphthalene (where M = alkali metal)	Effective catalyst	535
	NaCN in toluene or THF	Polymer obtained has the structure $-\!\!\left(CH_2-O\right)\!\!-_n$ with $CH=CH_2$ pendant on CH_2	535

	Li pyridine, Li quinoline, NaH in toluene or THF at -40° and $-78^{\circ}C$	Soluble polymers	536
Adipimide	Na adipimide	Triacetamide as initiator; polymer insoluble in monomer	537
Aldehydes (C_3–C_7)	*t*-BuOLi at low temperatures	Crystalline polymers	538
	s-BuOLi, $LiAlH_4$, BuLi in alkenes at -40 to $-80^{\circ}C$	Crystalline polymer	539
	$(EtO)_2Mg$, $(EtO)_2Ca$, BuLi, Et_2Zn, EtMgBr, $(EtO)_2Zn$ in hexane at $-78^{\circ}C$	Crystalline polymer	540
	Alkali metal alkoxides; alkali metals, alkali metal alkyls, Grignard reagents, $LiAlH_4$, at low temperatures	Crystalline isotactic polymers; polymerization mechanism	541, 542
	Benzophenone–Li in THF at $\leqslant -68^{\circ}C$	A state of equilibrium between the monomer and polymer observed in the range -68 to $-90^{\circ}C$	543
Aldehydes and ketones (α,β-unsaturated)			
Acrolein, methacrolein, crotonic aldehyde	K ketyl in polar solvents	Polymerization initiated mainly by electron transfer to monomer	544
Furane-substituted aldehydes and ketones	Na	Aldehydes give chiefly Na-containing polymer (90–95%). Ketones give only 9–17% fractions containing Na ion. H_2O soluble low MW polymer (590–1200)	545
Ketones	*n*-Bu_2Zn, *n*-Bu_3Al	Conjugate addition and hydrogen abstraction reactions occurred	546
	n-BuMgBr	*n*-Butane and conjugate addition products observed. Solvent effect was studied	547
β-Aminopropionitrile	*s*-BuONa in *o*-dichlorobenzene at $120^{\circ}C$	Polyamidine produced $\left(NH-CH_2CH_2-\underset{\underset{NH}{\Vert}}{C}\right)_n$	548

(*table continues*)

TABLE 2.2 Heterounsaturated Monomers (*continued*)

Monomer	Initiator system	Remarks	References
n-Butyraldehyde azine $CH{=}N{-}N{=}CH$ with $CH_2CH_2CH_3$ on each CH	BuLi in bulk or solution polymerization at −40 to +30°C	Formation of a methanol-insoluble oligomer (MW ~3000) in low yield (16.8–26.2%) with high yield of a methanol-soluble oligomer of MW ~ 1400	549
β-Cyanopropionaldehyde	Polystyryldisodium, Na naphthalene, disodium tetraphenylethylene	Low MW, an aldol condensation-type product for the last two initiators	550
	benzophenone–monosodium, -disodium, and -dilithium complexes	Low MW poly(cyanoethyl)oxymethylene with high stereoregularity	551
Ketenes			
Dialkyl (methyl, ethyl, butyl)	ΦLi or BuLi in toluene at −75 to 20°C	IR spectra showed the presence of ester group	552
Dimethyl	Grignard reagent in ether at −20 to 30°C	Oligomeric polyketone and nondistillable polyester formed at the same time; mechanism proposed	553
	Na naphthalene, NaCN, $K_2(C_6H_5)_2CO$, MgBrEt, $ZnEt_2$, LiBu–$ZnEt_2$, Et_3N	Three types of structures possible: polyketone, polyester, polyacetal depending on the type of initiator and polymerization conditions	554
	Et_3N, piperidine, Φ_3P in acetone at −70°C	Crystalline polymer having polyacetal structure	555
	ΦMgBr, $ZnCl_2$, *n*-BuLi	Both crystalline and amorphous polymer prepared	556
	BuLi, BuOLi in ether	Crystalline polymer with essentially polyacetal structure	557
	NaOMe in ether at room temperature	A polyester prepared	558
Formaldehyde	Et_3N, Bu_4N^+ laurate, calcium stearate in toluene	Kinetics of polymerization	559–562
	Lewis bases in ether, acetone, and toluene at −78°C	Basicity, steric, and structural influences of the Lewis bases are the important factors for determining their reactivity as initiators	563

	R_3N, R_3P	Mechanism of polymerization	564
	Et_3N	Anionic mechanism proposed	565
Furfural	Li, Na, K, and their polycyclic aromatic complexes	Highest yield (about 68.5%) was obtained with K in benzene. Polymer soluble in polar solvents	566
β-Methoxypropionaldehyde	Benzophenone–monolithium complex in THF at −90 to −70°C	Equilibrium between polymerization and depolymerization	567
o-Phenylenediacetaldehyde	*t*-BuOLi	Polyacetal 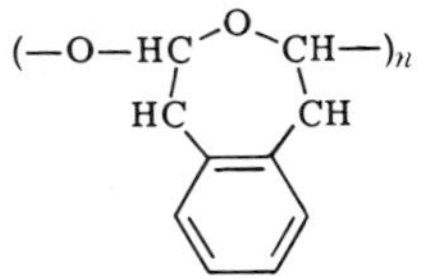	568
o-Phthalaldehyde	*t*-BuOLi in THF or toluene at −78°C, Na naphthalene, Na benzophenone	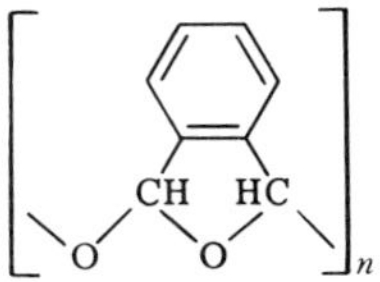 60% cis	569
N-Phenylmaleimide	NaCN, NaOH in DMF at 0°C	Polymerization occurs	570
Propiolaldehyde CH≡C−CH=O	NaCN or pyridine in THF at −78°C	Poly(ethynyloxymethylene) $[-CH(-C\equiv CH)-O-]_n$	571
	NaCN, *n*-Bu_3P in DMSO at −40°C	Colored polymers	535
Succinonitrile	NaOMe	Polymer with the structure 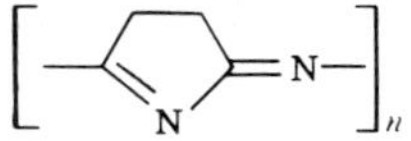	572

(*table continues*)

TABLE 2.2 Heterounsaturated Monomers (*continued*)

Monomer	Initiator system	Remarks	References
Succinonitrile (*continued*)	*t*-BuOK	Polymer with the structure 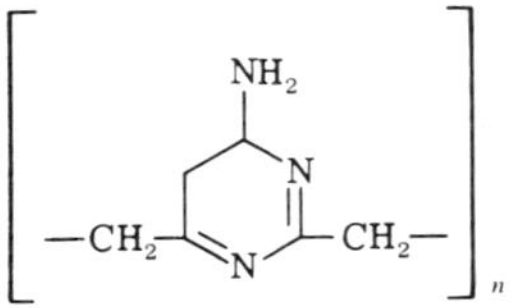	
	Et_2NH at 70°C	A brown-black polymer	573
	NaOMe in MeOH	Extremely slow reaction; yields a black polymer	574
Isocyanates			
Monoisocyanates	Na, NaCN, Na naphthalene, Na benzophenone ketyl in polar solvents at −20 to −100°C	Formation of Nylon-1 and cyclic dimer and trimer	575, 576
n-Butyl and phenyl	EtLi, *n*-octylsodium, EtOLi, Et_2Be in CS_2 or THF at −78°C	Crystalline polymers	577
	Diethyl sodioethylmalonate, ethyl sodioethylacetonate, and sodioethylacetylacetone in toluene at −75°C	Low yield	578
	NaCN in $HCONMe_2$	Optically active polyisocyanates	579, 580
n-Butyl	*n*-BuLi in toluene at −78°C	Rodlike polymer	581
	n-BuLi in toluene at −78°C	Polymers with narrow MWD, $\overline{M}_w/\overline{M}_n \approx 1.1$	582
n-Butyl, *n*-pentyl, phenyl, and cyclohexyl	NaCN or BuLi in DMF	Crystalline polymers	583
n-Hexyl	*n*-BuLi in hexane	High molecular weight; polymers are rodlike	584
	NaCN in DMR	High molecular weight	585
n-Octyl	*n*-BuLi in hexane	High molecular weight; polymers are rodlike	581, 584

Cyanoformamidyl	Et_3N in DMF	Rapid reaction to yield polymer containing the following structural unit 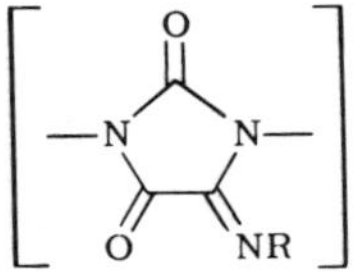	586
Vinyl isocyanate (see ethylene and vinyls)			
Diisocyanates			
Aliphatic 1,2-diisocyanates	NaCN in DMF at -30 to -50°C	Cyclopolymerization to linear high MW product having regularly recurring 5-atom heterocyclic units	587
cis-1,2-Cyclopropane diisocyanate	NaCN in DMF at -40°C	Polymer with the structure 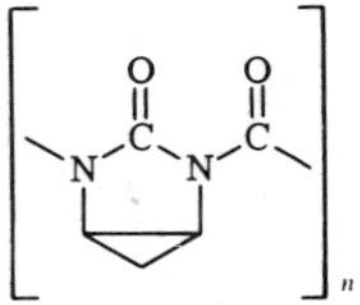 prepared	588
1,3-Diisocyanatocyclohexane	NaCN in DMF at -78°C	Cis isomer polymerizes to give soluble polymers containing little or no residual isocyanato groups; no polymerization occurs for trans isomers	589, 590
Carbodiimides			
Diethyl, di-*n*-butyl, di-*n*-hexyl, diphenyl, diallyl, methylisopropyl	*n*-BuLi in toluene at 25°C	The chain repeat unit is 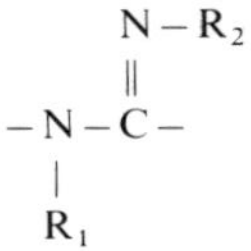	591

TABLE 2.3 Cyclic Monomers (Heterocyclics)

Monomer	Initiator system	Remarks	References
Oxides			
Ethylene oxide	Na, K catalysts in $(Me_2N)_3PO$	Relation between $[\eta]_{living}$ and $[\eta]_{deactivated}$ was calculated and compared with experimental result	592
	Na, K, or Cs naphthalenes in THF	Polymerization kinetics	593
	K^+, Cs^+, Na^+ in Me_2SO	K^+ and Cs^+ chain growth only via free ion; Na^+ ion pairs also participated in the reaction; chain transfer to solvent	594
	Li naphthalene, K naphthalene, K biphenyl, K anthracene, or di-K anthracene in Me_2SO or THF	No propagation occurred with Li naphthalene; polymers obtained for the other initiators	595
	Na, K, and Cs naphthalene in THF	Strong association; living systems	596
	Na and K alcoholates of the monomethyl ether of diethylene glycol in $(Me_2N)_3PO$	Association presence; reaction rate depressed by $KB\Phi_4$	597
	t-BuOK in Me_2SO	Polymerization kinetics. It is believed that initiation occurred via reaction of $MeSOCH_2^-$ with monomer	598
	Na naphthalene in THF	Initiation was by direct addition of the monomer to Na naphthalene, and the resulting polymer contained dihydronaphthalene residue	599
	t-BuOK in Me_2SO	No chain transfer reaction	600
	BuONa		601
	MeOK, MeONa, MeOLi	Lithium and sodium methoxide showed little catalytic activity	602
	Na alcoholate/alcohol	Optimum condition for the formation of high polymer was salt/alcohol = 1	603
	K 4-(phenylazo)benzyl alcoholate	UV-photometric determination of the MW with different end groups	604
	Cs 3,6-dioxa-1-octanolate in Me_2SO at 50°C	Conductivity measurements	605

	KOH (anhydrous) in THF	Living system; low catalyst efficiency; rate increased by Me_2SO	606
	9-Methylfluorenyl alkali salts and fluorenyl alkali salts in polar solvents at −30°C	Same behavior as that of the corresponding fluorenyl salts in anionic polymerization; spectrophotometric study of the kinetics	607
	$Cs(OCH_2CH_2OCH_2CH_2OEt)$ in $(Me_2N)_3PO$ at 40°C	Kinetics of polymerization; rate depressed by $CsB\Phi_4$	608
	K alcoholate of 4-(phenylazo)benzyl alcohol	Incorporation of initiator M_n from UV is not the same as M_v	609
	Salts of living polystyrene, α-methylstyrene tetramer, polytolan, carbazole in THF	Kinetics of polymerization	610
Propylene α-D-oxide	KOH, Et_2Zn–H_2O, bulk or benzene	NMR study of the polymer microstructures	610a
cis-Propylene β-D-oxide	Et_2Zn–H_2O, toluene or bulk	High yield (> 80%)	610b
trans-Propylene β-D-oxide	*t*-BuOK, toluene or bulk	Reaction mechanism elucidated by NMR spectroscopy	
Propylene oxide	MeOLi, MeOK	No polymerization for MeOLi due to strong association in $(Me_2N)_3PO$; DMR and $(Me_2N)_3PO$ accelerated the reaction and MeOH in small amounts retarded it	611
	Et_2Zn	N-compounds such as phenothiazine, *m*-aminophenol, piperidine are added as cocatalysts	612
	Et_2Zn	O_2, H_2O, alcohols were employed as cocatalysts to yield high polymers	613
	KOH (anhydrous)	Polymer containing terminal unsaturation	614
	t-BuOK in Me_2SO	Chain transfer reaction	600
	KOH	Low MW polymers	615
	$FeCl_3$	High MW isotactic polymers	615
	Potassium monoglycerate	Reaction rate determination	616
	t-BuOK, Na naphthalene at 20–60°C in Me_2SO or THF or Me_2SO–THF mixtures	Reaction orders	617
	t-BuOK in Me_2SO	Low MW polymer with unsaturated end groups	240

(*table continues*)

TABLE 2.3 Heterocyclic Monomers (*continued*)

Monomer	Initiator system	Remarks	References
Propylene oxide (*continued*)	Zn salts of xanthates, dithiocarbamates and thiocarboxylates	High polymers	618
	t-BuN(EtZn)$_2$ + H_2O (trace) (bulk)	Isotactic polymer, addition of water enhances rate and also yield of isotactic	619
1,2-Butylene oxide	*t*-BuOK or dimsylsodium at 30–60°C in Me_2SO and Me_2SO–THF mixtures	Reaction mechanism	620
	Zn salts of xanthates, dithiocarbamates, and thiocarboxylate	High polymers	618
t-Butyl oxirane	$ZnEt_2$–H_2O	Structural studies of the isotactic polymer by X-ray diffraction method	621
	$ZnEt_2$–H_2O, 65°C, bulk; *t*-BuOK at 96°C	Isotactic semicrystalline polymer; thermal transition study by DSC	622
	t-BuOK in bulk or $Zn(OCH_3)_2$ in benzene	High yield; the MeOH-insoluble fraction has isotactic structure	623
2,3-Dideuteromethyloxirane	$(RO)_2Zn$ in benzene at 35°C	Stereochemistry of polymerization based on NMR	624
Trifluoromethyloxirane	KOH	Low MW polymer (970–4300) containing terminal unsaturation	625
	Monosodium salt of hexafluoropentanediol	Polymer terminated with primary hydroxyl groups	625
Epichlorohydrin	NaOH	Acetylene and hydrogen were generated through the polymerization	626
Optically active 2-methylbutyl-glycidyl ether	KOH in hexane	The concentration of KOH did not affect the polymer yields	627
Optically active methyl glycidyl ether	KOH	Optically active polymer	628
Isoamyl glycidyl ether	Et_2Zn/H_2O or KOH	Optically active polymer	629
Styrene oxide	MeOK at 80°C	Chain transfer reaction involving the C–H bond of the epoxide ring occurs	630
	Potassium aromatic nitriles or ketones in THF (65–80°C)	Initiation mechanism depends both on the monomer and the metal complex used	631

	Zn salts of xanthates, dithiocarbamates, thiocarboxylate	High polymer	601
Sulfides			
Ethylene sulfide	Et_2Zn–I_2 in THF	Polymer was stabilized against oxidation by the ethylene sulfide–halogen adduct formed during polymerization reaction	632
	Na naphthalene in THF at −70° and 0°C	The polymer does not contain naphthalene or dihydronaphthalene groups in the chain	633
Propylene sulfide	Carbazylsodium in THF at −40 to 0°C	Living polymer, kinetic study. Polymerization rate was slowed in the presence of sodium tetraphenylboride	634
	Quaternary ammonium salt in THF at −30°C	Living polymers, kinetic study	635
	Sodium thiolate in THF at −30°C	Thiolate group association	636
	KOH	Optically active monomer gave optically active poly(propylene sulfide)	637
	Na naphthalene in THF at −70° and 0°C	The polymer does not contain naphthalene or dihydronaphthalene group in the chain (cf. ethylene oxide)	633
	Na naphthalene in THF at −30°C	Living polymers with two stable active centers	638
	Na and K naphthalene	Determination of dissociation constants	639
	Φ_3CNa in THF	The MW determined was double the calculated value, possibly due to the homolytic rupture of the C−Na bond	640
	Na naphthalene or carbazylsodium in THF	Determination of rate constant; kinetics of polymerization	641
	EtLi, BuLi, Li ethanethiolate or 1,4-dilithio-1,1,4,4-tetraphenylbutane in aprotic solvents at −78°C	Essentially free of termination and chain transfer reaction	642
	Na naphthalene; carbazylcesium in THF	Kinetics of polymerization in the presence of macroheterobicycles	643

(*table continues*)

TABLE 2.3 Heterocyclic Monomers (*continued*)

Monomer	Initiator system	Remarks	References
Propylene sulfide (*continued*)	Na naphthalene, carbazylsodium, 9-methylfluorenylcesium, carbazylcesium in polar solvents	The influence of the solvent and macroheterobicycles on the kinetics of polymerization	644
	n-BuLi, LiOR* (* chiral group)	Optically active polymer; mechanism proposed	645
	$ZnEt_2$/3,3-dimethyl-1,2-diol in toluene at 25°C	Optically active polymer, up to 90% pure isomer prepared from racemic monomer mixture	646
Isopropylthiiranes (racemic and optically active)	Na, $ZnEt_2$–H_2O (1:1); $ZnEt_2$–3,3-dimethyl-1,2-butanediol (1:1) (bulk or toluene at 25°C)	Optically active polymers	647
Thietane (thiacyclobutane) and 2-methylthietane	RLi in THF	Living polymer, having carbanionic living end (not thiolate)	648
Lactones			
β-Propiolactone	$(CH_3)_3N^+-CH_2COO^-$ in ethanol at 25°C	Polymerization via zwitterion; chain transfer with the solvent occurs	649
α,α-Disubstituted *β*-propiolactones	K benzoate and Et_4N^+ benzoate in Me_2SO and THF	The solvent and counterion effect	650
	R_3N	Polymerization via zwitterion	651
	Tetrahexylammonium benzoate	Crystalline material	652
	Quaternary ammonium carboxylates in various solvents (acetonitrile, THF, H_2O)	Kinetics of polymerization	653
Optically active *α*-(*p*-substituted benzenesulfonamide) *β*-propiolactone	Et_3N, BuLi in bulk, ethyl acetate or dioxane	Rate of polymerization	654
α,α-Dialkyl *β*-lactones	Quaternary ammonium carboxylates (acetonitrile, THF, H_2O at 25 and 35°C)	Kinetic studies; polar substituents in the alkyl groups accelerate the rates, whereas branched alkyls or bridged rings retard	653
	Tetrahexylammonium benzoate in THF	DL monomer mixture gives an atactic microstructure	652

α,α-Spirobicyclic β-lactones	Phosphobetaine catalyst, Et_4N^+ pivalate, acetonitrile	High MW polyesters for cyclohexene and bicycloheptane monomers, lower MW for the less reactive bicycloheptene and ethaneanthracene monomers	655
Cyclobutane 1,3-lactones	Na, K, $NaOCH_3$, NaK, Φ_3CNa, *t*-BuOK, Φ_3COK, bulk or xylene at $T > 35°C$)	Readily polymerized; high MW, rather high melting polyesters; soluble in acidic solvents	656
Tetramethyl glycolide	*t*-BuOLi in bulk at 120–140°C	Yield >90% poly(tetramethyl glycolide); $\left[-O-\overset{CH_3}{\underset{CH_3}{C}}-\overset{O}{\overset{\Vert}{C}}- \right]$ MW ≤ 250,000	657
Lactams			
ε-Caprolactam	EtMgBr in Et_2O at 100–200°C	Aromatic or hydroaromatic carboxylic acid esters, such as dimethyl terephthalate, dimethyl *cis*-1,4-cyclohexanedicarboxylate, etc., were used as promoters	658
	LiH, $LiNH_2$	A promoter is formed *in situ* by the reaction of 200–1400 ppm water in the monomer with sufficient isocyanate to reduce water content to <50 ppm	659
	NaH, NaOMe at 145–175°C	1-Chloro-3-oxoisoindolenine as a promoter	660
	Na caprolactam salt at 120°C	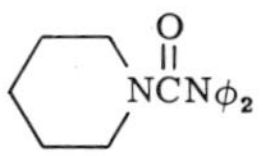 as polymerization accelerator. The polymer formed did not dissolve completely in 96% H_2SO_4	661
	NaH at 145–175°C	$\Phi N{=}CCl_2$ as a promoter	662

(*table continues*)

TABLE 2.3 Heterocyclic Monomers (*continued*)

Monomer	Initiator system	Remarks	References
ε-Caprolactam (*continued*)	Na	Accelerated by *N*-methacryloyl-ε-caprolactam or $CH_2{=}CH{-}CON\Phi_2$ (CH=CH bearing CONΦ₂); partial copolymerization of the accelerators occurs	663
	$LiAlH_4$	$(Me_2N)_3PO$ was added to the polymerization mixture and retained as a plasticizer for the polymer	664
	Na caprolactam salt	Addition of ~1 mole % of multivalent cations in the polymerization mixture decreased the crystallinity of the polymer	665
	Na in ClCN at 180°C	A solid polymer was produced	666
	BuLi at 170°C	*N*,*N*′-diphenyl-1,3-diazo-2,4-cyclobutanedione was used as a promoter	667
	Na caprolactam/*N*-acetylcaprolactam	The effect of catalyst/cocatalyst ratio on rate of polymerization	668
	NaH/1,3,5-trimethylbiuret at 150°C	Biuret is added as a cocatalyst	669
	Na dihydrobis(methoxyethoxy) aluminate/*N*-benzoylcaprolactam	Side reactions result in a bimodal MWD of the polymer	670
	NaH/substituted phosphoranes	Phosphoranes used as a cocatalyst	671
	NaH/*N*-benzoylcaprolactam	Effect of concentration of catalyst/cocatalyst and polymerization temperature on MW of polymer	672
	NaH at 160°C	Polymethylenepolyphenyl isocyanate (PAPI) as cocatalyst	673
	Na ε-caprolactam/*N*-acetylcaprolactam at 155°C	2,4-Diphenoxy-6-methylamino-*s*-triazine was used as an activator	674
	Sodium caprolactam/*N*-acetylcaprolactam (1:1)	Polymerization mechanism proposed	675
	NaH at 185°C	Triallylisocyanurate as a promoter	676
	NaH	2,4-Dimethylallophanoyl chloride as a cocatalyst	677
	LiH at 95°C	Triphenoxy-*s*-triazine as cocatalyst	678

	Na caprolactam salt at 120°C	Activating efficiencies of isocyanates were investigated	679
	NaH or LiH at low temperature	Activated by triallylisocyanurate	680
	NaH / 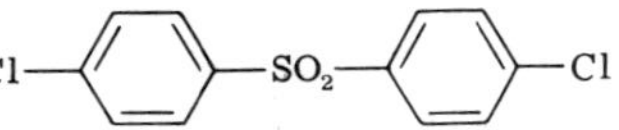	Other activated aromatic halides also investigated	681
	Na caprolactam/N-benzoylcaprolactam at 220–225°C	Effect of reaction conditions on the content of acidic and basic groups in the polymer	682
	NaH/*N*,*N*′-diisopropylcarbodiimide, *N*,*N*′-diphenylurea, 1,3-diethylurea, *n*-butyl acetate, phenyl isocyanate, acetylcaprolactam, benzoylcaprolactam	Influence of various cocatalysts	683
	Na caprolactam/*N*-carbanilinocaprolactam, bromomagnesium caprolactam/*N*-acetylcaprolactam, at 130°C	Kinetics of polymerization	684
α-Piperidone	Me_4N^+ piperidone	Formation of Nylon-5	685
	R_4N^+ α-piperidone salt	Impurities such as H_2O interfere with the chain transfer reaction	686
	LiCl or LiSCN	Low polymerizability due to the crystalline structure of the polymer	687
	Alkali metal and acyl compound	Formation of poly(5-valeramide) (maximum conversion ~76%)	688
	K, or $MAlEt_3$, or $MAlEt_4$ (where M is Li, Na, or K) at low temperatures	*N*-acyl-α-piperidone was used as initiator to give high MW	689
β-Propiolactam	K pyrrolidone in Me_2SO	*N*-acetylpyrrolidone as a coinitiator; kinetics of polymerization	690
3-Methyl-3-butyl-β-propiolactam	BuLi in toluene at 25°C	Soluble polymer	691
3,4-Disubstituted β-propiolactam	K pyrrolidone in halogenated benzene/petroleum oil mixtures	Fine polymer grains; 2-phase suspension polymerization	692
Optically active β-propiolactam	K pyrrolidone in Me_2SO	Polymers having varying optical purities	693
α-Pyrrolidone	Na pyrrolidonate or $R_4N^+X^-$	High MW Nylon-4	694

(*table continues*)

TABLE 2.3 Heterocyclic Monomers (*continued*)

Monomer	Initiator system	Remarks	References
α-Pyrrolidone (*continued*)	$R_4N^+ \cdot α$-pyrrolidone salt	Impurities interference	695
	KOH in $COS–CO_2$ mix	Polymer with polydispersity of 1.90–2.30	696
	t-BuOK/γ-butyrolactone	Effect of temperature and initiator/activator concentrations on the anionic polymerization	697
	R_4N^+ α-pyrrolidone salt	High MW polymers	698
	K α-pyrrolidone/CO_2 salt	Maximum yield at 50°C, but highest relative viscosity at ambient temperature	698
	K α-pyrrolidone salt/*N*-benzoylpyrrolidone	Maximum polymerization rate at 40°C	699
	Pyrrolidonate magnesium bromide in THF	*N*-Benzoyl-, *N*-acetyl-, and *N*-methylpyrrolidone as cocatalyst	700
	K pyrrolidonate	NMR study of the mechanism of polymerization	701
Epoxypropylpyrrolidone	KOH or Et_2NH	Yield depends on catalyst concentration	702
	KOH, Et_2NH, K pyrrolidone at 40–75°C	Oligomers with MW 500–2000	703
Epoxypyrrolidone	KOH, K pyrrolidone, Et_2NH, and K alcoholate	Polymer with MW ~3000	704
8-Oxa-6-azabicyclo[3.2.1]octan-7-one	NaH in THF (0°C) or neat (100°C)	Formation of polyamide 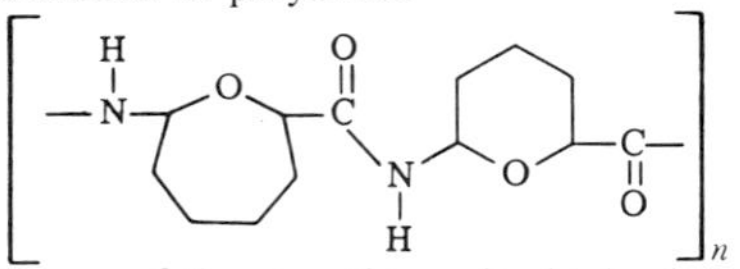	705
Aza-2-keto[2.3]paracyclophane	NaH	Polymer of the repeating unit obtained 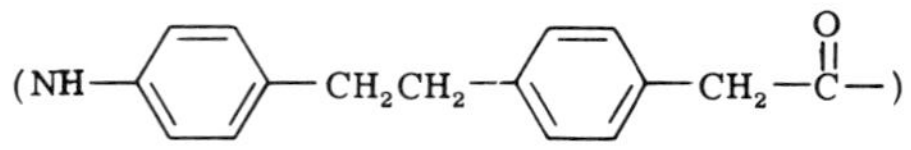	706
Anhydrides			
Maleic anhydride	$Φ_3P$, Bu_3P in Me_2SO	Proceeded by zwitterion mechanism	707

	Pyridine bases, Φ_3P, Bu_3P	Spontaneous polymerization in the presence of catalyst	708
	Pyridine, picoline in acetic anhydride	Kinetic and conductivity measurement suggests that the polymerization rate depends on the basicity of the initiator	709
	Pyridine	Formation of dark polymers	710
	Et_3N, pyridine	Dark polymer displaying the characteristic properties of polyconjugated polymers; soluble in polar solvent	711
	Pyridine	Slow reaction; oligomeric product	712
N-Carboxy anhydrides	$BuNH_2$ in MeCN	Insoluble polymer with active amino end group which can initiate the polymerization of NCA in MeCN; block copolypeptides were synthesized	713
	Primary amines (diamines) in acetonitrile	Crystalline polyglycine	714
	Diamines in acetonitrile	Wider MWD than those prepared by $BuNH_2$; crystalline polymers	715
	Pyridine, γ-picoline, Et_3N, *t*-$BuNH_2$	Polymerization mechanism	716
	Primary amines	The polymerization characteristics studied	717
	α,ω-primary diamines in $HCONMe_2$	Kinetics of polymerization	718
	Et_3N	Polymerization mechanism	719
	$BuNH_2$ in acetonitrile	Effect of the steric hindrance on the crystal growth by the side chain	720
	$BuNH_2$, Et_3N, NH_3, and H_2O in acetonitrile, Me_2SO, nitrobenzene	Conformation of polymers	721
	Benzylamine, *n*-hexylamine, Et_3N, MeONa, Na *N*-benzylcarbamate	Elucidation of the mechanism of polymerization using ^{14}C-labeled initiators	722
S-Carboxy-α-mercaptoacetic anhydride	Amines, NaOMe, H_2O or $NaBH_4$ in bulk or solution (THF, dioxane) polymerization at -50 to $+200°C$	Poly(thioglycolide) $\left(S-CH_2-\underset{\underset{O}{\parallel}}{C} \right)_n$ with MW $< 36{,}000$	723
N-Thiocarboxy anhydride (NTA)	Primary amine, tertiary amine	Give peptide with lower yield and DP than does that of the corresponding *N*-carboxy anhydride	724

(*table continues*)

TABLE 2.3 Heterocyclic Monomers (*continued*)

Monomer	Initiator system	Remarks	References
N-Thiocarboxy anhydride (NTA) (*continued*)	Pyridine, γ-picoline, Et_3N, *t*-$BuNH_2$	Polymerization mechanism	716
Siloxanes and Other Silicon Cyclics			
Octamethylcyclotetrasiloxane	KOH, CsOH, RbOH, LiOH, NaOH, K_2O, MeOK, MeONa, KNH_2	The base strength of the catalyst is the rate-controlling factor	725
	KOH	Kinetics of polymerization	726
	Di-K polydimethylsiloxane diolate at 140°C	Polydispersity of the elastomer studied	727
	KOH, Na, K, Na and K naphthalene in THF	Catalytic effect of THF	728
	KOH, K silanolate, MeOK, in decalin, *p*-chlorotoluene, *o*-dichlorobenzene, and nitrobenzene	Effect of solvent polarity on the base-catalyzed polymerization	729
	K siloxanediolate 110 to 165°C	Chromatographic study of the cyclic molecules present in an equilibrium polymerization	730
	KOH	Effect of small amount (0.01%–1%) of DMSO on the base-catalyzed polymerization and equilibration of methylsiloxanes	731
	K silanolate	The catalytic effect of $(Me_2N)_3PO$	732
	K polysiloxanediolate at 140 ± 5°C	MW and MWD modulating	733
	Dodecylbenzenesulfonic acid at 50 to 100°C	The MW of the polymer is controlled by the temperature of polymerization; anionic emulsion polymerization	734
Cyclic polydimethylsiloxanes (3–9 silicon atoms)	KOH, NaOH, LiOH	The overall rate of polymerization differs markedly for different monomers and catalysts, though the activation energy is the same	735
Hexamethylcyclotrisiloxanes	Biscatecholsiliconate; BuLi, Li silanolate; Li siloxanolate	Monodisperse polymers	736

	Benzyltrimethylammonium bis(*o*-phenylenedioxy)phenyl siliconate/Me_2SO–H_2O	Linear difunctional polymers	737
	s-BuLi	MWD broadened at very high MWs due to side reaction	738
	$(CH_3)_2Si(OLi)_2$ in hexane–$(Me_2N)_3PO$ at 25°C	Polymers of MW ranging from 2×10^3 to 2×10^6 with narrow MWD	739
	K dimethylphenylsilanolate	Kinetics of initiation was studied	740
Hexaorganocyclotrisiloxanes	Li dimethylsilanolate, Li triorganosilanolate; organolithium, LiOH, Li alkoxide, LiH, and $LiAlH_4$ in electron-donating aprotic solvents	Homopolymers and block copolymers have been reported	741
1,1,3,3,5-Pentamethyl-5[*m*-(trifluoromethyl)phenyl]cyclotrisiloxane	Na methylphenylsiloxanediol	Reaction kinetics	742
1,3,5-Trimethyl-1,3,5-tris-(3,3,3-trifluoropropyl)cyclotrisiloxane	Na, K, or Li salts of 1,3,5-trimethyl-1,3,5-tri(3,3,3-trifluoropropyl)trisiloxanediol, or 1:2 mixtures of H_2O with Bu_2NH, piperidine, Et_3N	The rate of polymerization depended on the nature of the catalyst counterion and decreased in the order: $Na^+ > Li^+ > K^+ \gg Bu_2N^+H_2 >$ piperidinum ion $> Et_3NH^+$	743
Cyclotetrasiloxanes with cyanoalkyl and methyl groups	K polydimethylsiloxanolate, Na trimethylsilanolate, Li trimethylsilanolate	The presence of $(CH_2)_2CN$ or $(CH_2)_3CN$ groups in the monomer increase the rate constants of the polymerization	744
1,1,3,3-Tetramethyl-5,5,7,7-tetraphenylcyclotetrasiloxane	$Me_4N(SiMe_2O)_nNMe_4$	Polymerization process is accompanied by intramolecular chain transfer reaction already in the initial stage of polymerization	745
Methylpropyldimethylcyclosiloxanes	K and Na siloxanediolate in bulk or acetone solution polymerization	The rate of polymerization decreases with increase in the number of methylpropylsiloxane units in the ring	746
Methylarylcyclosiloxanes	K poly(dimethylsiloxane)diolate at 120°C	Polymerization kinetics	747
Methyl-(10-carbomethoxydecyl)cyclosiloxanes	K naphthalenes in THF	Equilibrium polymerization; low MW polymers due to steric effects of the bulky long-chain substituents	748

(*table continues*)

TABLE 2.3 Heterocyclic Monomers (*continued*)

Monomer	Initiator system	Remarks	References
Cyclosiloxanes with 3,3,3-trifluoropropyl groups on silicon atom	K, Na, and Li hydroxides or siloxanolates	An increase of rate of polymerization due to the I-effect of fluorine atoms	749
Bis(dimethyldiphenyldisiloxane)-spirosilane	α-Hydroxy-ω-tetramethylammoniumhydroxymethylphenylsiloxane	Polymerization rate and activation energy determination	750
Organospirosiloxanes	$Me_4NO(SiMe_2O)_{26}NMe_4$	Soluble polymers with some degree of gel	751
1,1-Dimethyl-1-silacyclobutane	Me_3SiONa, Me_3SiOK, NaOH, or KOH	Reaction via ring cleavage to yield a linear polymer containing C – Me bonds	752
Hexa(trimethylsilyloxy) cyclotrisiloxane	BuLi	Low MW linear polysilicates of narrow MWD	753
5-Membered cyclocarbosiloxanes	Li *n*-butyldiphenylsilanolate in THF	Effect of temperature, concentration of THF, initiator, and water on the rate of polymerization, MW, and MWD of polymer	754
	K dimethylsilanolate at 120–130°C in bulk or solution polymerization (Φ_2O)	High polymer	755
	Na phenyldimethylsilanolate in non-polar media	No redistribution of siloxane units between living ends of the polymer	756

Miscellaneous Heterocyclics			
Bicyclo[2.1.0]pentane-1-carboxylate, bicyclo[2.1.0]pentane-1-carbonitrile, bicyclo[3.1.0]hexane-1-carbonitrile	*n*-BuLi, Et_2Mg, CH_3Li, THF, toluene at −80°C	Low MW polymer for the carboxylates and a high yield of polymer from the nitrile; no polymers were obtained for the bicyclo[3.1.0]hexane-1-carbonitrile	757
Cyclic diamides	$NaOCH_3$, KOH, LiH, CaH_2, Na, Φ_3P, diamine, acetates of Group I and II metals at 259°C	Formation of polyoxamides; polymerization occurs rapidly in the melt; polymerization also occurs in the crystalline state	758
Propylene carbonate	Et_2Zn at 120–180°C	A pale yellow, viscous material of relatively low MW (1000–4000)	759
1,3-Propylene methyl phosphate	EtONa; *t*-BuOK in diglyme or Me_2SO	Reversible polymerization	760

References

1. D. C. Pepper, *Q. Rev., Chem. Soc.* **8**, 88 (1954).
2. M. Szwarc, *Fortschr. Hochpolym.-Forsch.* **2**, 275 (1960).
3. J. E. Mulvaney, C. G. Overberger, and A. M. Schiller, *Fortschr. Hochpolym.-Forsch.* **3**, 106 (1961).
4. C. E. H. Bawn and A. Ledwith, *Q. Rev., Chem. Soc.* **16**, 361 (1962).
5. M. Morton, *in* "Copolymerization" (G. Ham, ed.), p. 421. Wiley (Interscience), New York, 1964.
6. S. Bywater, *Fortschr. Hochpolym.-Forsch.* **4**, 66 (1965).
7. M. Morton and L. J. Fetters, *Macromol. Rev.* **2**, 71 (1967).
8. M. Morton, *MTP Int. Rev. Sci.: Phys. Chem., Ser. One* **8**, 1 (1972).
9. M. Morton and L. J. Fetters, *Rubber Chem. Technol.* **48**, 359 (1975).

9a. M. Morton and L. J. Fetters, *in* "The Stereo Rubbers" (W. Saltman, ed.), Chapter 5, p. 212. Wiley, New York, 1977.

10. M. Szwarc, M. Levy, and R. Milkovich, *J. Am. Chem. Soc.* **78**, 2656 (1956); M. Szwarc, *Nature (London)* **178**, 1168 (1956).
11. F. Wenger, *Makromol. Chem.* **64**, 151 (1963).
12. M. Morton, R. Milkovich, D. McIntyre, and L. J. Bradley, *J. Polym. Sci., Part A* **1**, 443 (1963).
13. R. Waack, A. Rembaum, J. D. Coobes, and M. Szwarc, *J. Am. Chem. Soc.* **79**, 2026 (1957).
14. F. M. Bower and H. W. McCormick, *J. Polym. Sci., Part A* **1**, 1749 (1963)
15. J. M. G. Cowie, D. J. Worsfold, and S. Bywater, *Trans. Faraday Soc.* **57**, 705 (1961).
16. H. W. McCormick, *J. Polym. Sci.* **36**, 341 (1959).
17. T. Lyssy, *Helv. Chim. Acta* **42**, 2245 (1959).
18. C. Geacintov, J. Smid, and M. Szwarc, *J. Am. Chem. Soc.* **84**, 2508 (1962).
19. D. N. Bhattacharyya, C. L. Lee, J. Smid, and M. Szwarc, *Polymer* **5**, 54 (1964).
20. D. N. Bhattacharyya, C. L. Lee, J. Smid, and M. Szwarc, *J. Phys. Chem.* **69**, 612 (1965).
21. G. Meyerhoff, *Z. Elektrochem.* **61**, 1249 (1957).
22. G. Meyerhoff, *Z. Phys. Chem. (Wiesbaden)* [N. S.] **23**, 100 (1960).
23. F. S. Dainton, D. M. Wiles, and A. N. Wright, *J. Polym. Sci.* **45**, 111 (1960).
24. C. Geacintov, J. Smid, and M. Szwarc, *J. Am. Chem. Soc.* **83**, 1253 (1961).
25. F. Wenger, *J. Polym. Sci.* **58**, 452 (1962).
26. F. Wenger, *J. Polym. Sci.* **60**, 99 (1962).
27. H. Hostalka and G. V. Schulz, *J. Polym. Sci., Part B* **3**, 175 (1965).
28. H. Hostalka, R. V. Figini, and G. V. Schulz, *Makromol. Chem.* **71**, 198 (1964).
29. G. Lohr and G. V. Schulz, *Makromol. Chem.* **77**, 240 (1964).
30. D. J. Worsfold and S. Bywater, *J. Chem. Soc.* p. 5234 (1960).
31. H. Hostalka and G. V. Schulz, *Z. Phys. Chem. (Wiesbaden)* [N. S.] **45**, 286 (1965).
32. D. Lipkin, D. E. Paul, J. Townsend, and S. Weissman, *Science* **117**, 534 (1953).
33. R. V. Figini, G. Lohr, and G. V. Schulz, *J. Polym. Sci., Part B* **3**, 985 (1965).
34. A. A. Korotkov and A. F. Podolsky, *J. Polym. Sci., Part B* **3**, 901 (1965).
35. N. D. Scott, *J. Am. Chem. Soc.* **58**, 2442 (1936).
36. F. Wenger, *Makromol. Chem.* **36**, 200 (1960).
37. S. N. Khanna, M. Levy, and M. Szwarc, *Trans. Faraday Soc.* **58**, 747 (1962); p. 1821 (with R. Asami).
38. A. A. Arest-Yakubovich, A. R. Gantmakher, and S. S. Medvedev, *Dokl. Akad. Nauk SSSR* **139**, 1351 (1961).
39. J. Stearne, J. Smid, and M. Szwarc, *Trans. Faraday Soc.* **62**, 672 (1966).
40. D. N. Bhattacharyya, J. Smid, and M. Szwarc, *J. Phys. Chem.* **69**, 624 (1965).
41. G. Allen, G. Gee, and C. Stretch, *J. Polym. Sci.* **48**, 189 (1960).

42. C. Stretch and G. Allen, *Polymer* **2**, 151 (1961).
43. S. P. S. Yen, *Makromol. Chem.* **81**, 152 (1965).
44. A. Zilkha, M. Albeck, and M. Frankel, *J. Chem. Educ.* **35**, 345 (1958).
45. F. J. Welch, *J. Am. Chem. Soc.* **81**, 1345 (1959); **82**, 6000 (1960).
46. K. F. O'Driscoll and A. V. Tobolsky, *J. Polym. Sci.* **35**, 259 (1959).
47. D. J. Worsfold and S. Bywater, *Can. J. Chem.* **38**, 1891 (1960).
48. S. Bywater and D. J. Worsfold, *Can. J. Chem.* **40**, 1564 (1962).
49. M. Morton, E. E. Bostick, and R. Livigni, *Rubber Plast. Age* **42**, 397 (1961).
50. S. Bywater and D. J. Worsfold, *J. Phys. Chem.* **70**, 162 (1966).
51. M. Morton, L. J. Fetters, and E. E. Bostick, *J. Polym. Sci., Part C* **1**, 311 (1963).
52. D. J. Worsfold and S. Bywater, *Makromol. Chem.* **65**, 245 (1963).
53. G. C. East, P. F. Lynch, and D. Margerison, *Polymer* **4**, 139 (1963).
54. R. C. P. Cubbon and D. Margerison, *Proc. Chem. Soc., London* p. 146 (1960).
55. Yu. L. Spirin, A. R. Gantmakher, and S. S. Medvedev, *Dokl. Akad. Nauk SSSR* **146**, 368 (1962).
56. Yu. L. Spirin, D. K. Polyakov, A. R. Gantmakher, and S. S. Medvedev, *Dokl. Akad. Nauk SSSR* **139**, 899 (1961).
57. R. C. P. Cubbon and D. Margerison, *Polymer* **6**, 102 (1965).
58. R. C. P. Cubbon and D. Margerison, *Proc. R. Soc. London, Ser. A* **268**, 260 (1962).
59. H. Hsieh, *J. Polym. Sci., Part A* **3**, 153, 173 (1965).
60. K. F. O'Driscoll, E. N. Ricchezza, and J. E. Clark, *J. Polym. Sci., Part A* **3**, 3241 (1965).
61. T. A. Altares, Jr., D. P. Wyman, and V. R. Allen, *J. Polym. Sci., Part A* **2**, 4533 (1964).
62. R. Waack and M. A. Doran, *Polymer* **2**, 365 (1961).
63. K. Kuwata, *Bull. Chem. Soc. Jpn.* **33**, 1091 (1960).
64. A. Zilkha, S. Barzakay, and A. Ottolenghi, *J. Polym. Sci., Part A* **1**, 1813 (1963).
65. H. Sinn, C. Lundborg, and K. Kirschner, *Angew. Chem.* **70**, 744 (1958).
66. W. Kern, D. Braun, and M. Herner, *Makromol. Chem.* **28**, 60 (1958); **36**, 232 (1960).
67. D. Braun, W. Betz, and W. Kern, *Makromol. Chem.* **42**, 89 (1960).
68. R. J. Kern, *Nature (London)* **187**, 410 (1960); *Polym. Prepr., Am. Chem. Soc., Div. Polym. Chem.* **4**, No. 2, 324 (1963).
69. J. L. R. Williams, T. M. Laakso, and W. J. Dulmage, *J. Org. Chem.* **23**, 638 (1958).
70. M. G. Evans, W. C. E. Higginson, and N. S. Wooding, *Rect. Trav. Chim. Pays-Bas* **68**, 1069 (1949).
71. W. C. E. Higginson and N. S. Wooding, *J. Chem. Soc.* p. 760 (1952).
72. N. S. Wooding and W. C. E. Higginson, *J. Chem. Soc.* pp. 774, 1178 (1952).
73. J. J. Sanderson and C. R. Hauser, *J. Am. Chem. Soc.* **71**, 1595 (1949).
74. J. E. L. Roovers and S. Bywater, *Trans. Faraday Soc.* **62**, 701 (1966).
75. G. Spach, M. Levy, and M. Szwarc, *J. Chem. Soc.* p. 355 (1962).
76. M. Levy, M. Szwarc, S. Bywater, and D. J. Worsfold, *Polymer* **1**, 515 (1960).
77. W. Schlenk, J. Appenrodt, A. Michael, and A. Thal, *Ber. Dtsch. Chem. Ges.* **47**, 473 (1914).
78. D. P. Wyman and T. Altares, Jr., *Makromol. Chem.* **72**, 68 (1964).
79. A. G. Kharitonov, V. S. Glukhovskoi, and B. I. Mikhant'ev, Deposited Doc., VINITI p. 320 (1974), *CA*, **86**:171921f.
80. J. Kralicek, V. Kubanek, J. Kondelikova, B. Casensky, and J. Machacek, Czech Patent 164,417 (1976); *CA* **85**, 33712s.
81. S. Bywater and I. J. Alexander, *J. Polym. Sci., Part A-1* **6**, 3407 (1968).
82. T. Shimomura, T. Smid, and M. Szwarc, *Polym. Prepr. Am. Chem. Soc., Div. Polym. Chem.* **8**, 879 (1967).
83. B. J. Schmitt and G. V. Schulz, *Makromol. Chem.* **121**, 184 (1969).
84. A. A. Korotkov and A. F. Podolsky, *J. Polym. Sci., Part B* **7**, 85 (1969).
85. N. Ise, H. Hirohara, T. Makino, K. Takaya, and M. Nakayama, *J. Phys. Chem.* **74**, 606 (1970).

86. J. Geerts, M. van Beylan, and G. Smets, *J. Polym. Sci., Part A-1* **7**, 2805 (1969).
87. H. Hirohara, M. Nakaya, K. Takaya, and N. Ise, *Trans. Faraday Soc.* **66**, 1165, 3163 (1970).
88. M. Tomoi and H. Kakinchi, *Kogyo Kagaku Zasshi* **73**, 2367 (1970); *CA* **74**, 54222v.
89. R. Plepys and R. Waack, U. S. Patent 3,560,469 (1971); *CA* **74**, 88384q.
90. R. L. Williams, D. H. Richard, and D. A. Salter, British Patent 1,211,193 (1970); *CA* **74**, 13572r.
91. G. E. Molau and J. E. Mason, *J. Polym. Sci., Polym. Chem. Ed.* **4**, 2336 (1966).
92. S. Bywater and D. J. Worsfold, *J. Organomet. Chem.* **10**, 1 (1967).
93. Y. Minoura and T. Hirahara, *Macromolecules* **8**, 42 (1975).
94. J. E. L. Roovers and S. Bywater, *Macromolecules* **8**, 251 (1975).
95. B. DeGroof, M. van Beylan, and M. Szwarc, *Macromolecules* **8**, 396 (1975).
96. B. DeGroof, W. Mortier, M. van Beylan, and M. Szwarc, *Macromolecules* **10**, 598 (1977).
97. F. C. Schwab, U. S. Patent 3,770,712 (1973); *CA* **80**, 83917e.
98. K. Takaya and N. Ise, *Polymer* **15**, 635 (1974).
99. K. Takaya, H. Tatsuta, and N. Ise, *Polymer* **15**, 631 (1974).
100. T. Narita, Y. Kunitake, and T. Tsuruta, *Makromol. Chem.* **176**, 3371 (1975).
101. T. Narita, T. Yasumura, and T. Tsuruta, *Polym. J.* **4**, 421 (1973).
102. C. Mathis, L. Christman-Lamande, and B. François, *Makromol. Chem.* **176**, 931 (1975).
103. G. Pizzirani, M. DiMaina, M. Palla, and P. Giusti, *Eur. Polym. J.* **13**, 605 (1977).
104. M. Bunge, G. Lohr, H. Hocker, and G. V. Schulz, *Eur. Polym. J.* **13**, 283 (1977).
105. A. L. Gatzke and E. Vanzo, *Chem. Commun.* p. 1180 (1967).
106. T. Shimomura, J. Smid, and M. Szwarc, *Makromol. Chem.* **108**, 288 (1967).
107. R. V. Figini, *J. Polym. Sci., Part C* **16**, 2049 (1967); *Makromol. Chem.* **107**, 170 (1967).
108. I. Sakurada, N. Ise, H. Hirohara, and T. Makino, *J. Phys. Chem.* **71**, 3711 (1967).
109. G. Lohr and G. V. Schulz, *Eur. Polym. J.* **11**, 259 (1975).
110. L. L. Bohm and G. V. Schulz, *Eur. Polym. J.* **11**, 265 (1975).
111. B. J. Schmitt and G. V. Schulz, *Eur. Polym. J.* **11**, 119 (1975).
112. J. Lacoste, S. Schue, S. Bywater, and B. Kaempf, *J. Polym. Sci., Polym. Lett. Ed.* **14**, 201 (1976).
113. S. A. Hurley and P. J. T. Tait, *J. Polym. Sci., Polym. Chem. Ed.* **14**, 1565 (1976).
114. K. A. Allen, B. G. Gowenlock, and W. E. Lindsell, *J. Polym. Sci., Polym. Chem. Ed.* **12**, 1131 (1974).
115. B. W. Brooks, *Chem. Commun.* p. 68 (1967).
116. T. Yamaguchi and Y. Hata, *Kobunshi Kagaku* **24**, 209 (1967).
117. B. Leuresse, E. Franta, and P. Rempp, *Eur. Polym. J.* **7**, 863 (1971).
118. H. Hirohara, M. Nakaya, and N. Ise, *J. Chem. Soc., Faraday Trans. 1* **68**, 58 (1972).
119. J. Komiyama, L. L. Boehm, and G. V. Schulz, *Makromol. Chem.* **148**, 297 (1971).
120. L. L. Boehm and G. V. Schulz, *Makromol. Chem.* **153**, 5 (1972).
121. I. M. Panaiotov and I. B. Rashkov, *J. Polym. Sci., Polym. Chem. Ed.* **10**, 1267 (1972).
122. G. R. Polyvakova and N. A. Pranikova, *Vysokomol. Soedin., Ser. A* **15**, 128 (1973); *Polym. Sci. USSR (Engl. Transl.)* **15**, 661 (1973).
123. M. Nashiwa, H. Yamamoto, and M. Maehara, Japan Kokai 72/34, 780 (1972); *CA* **78**, 137023q.
124. C. Roth, E. Giess, and K. Gehrke, *Plaste Kautsch.* **20**, 333 (1973).
125. K. Gehrke and C. Roth, *Plaste Kautsch.* **20**, 493 (1973).
126. H. Gisser and H. E. Mertwoy, *Macromolecules* **2**, 461 (1969).
127. K. S. Dennis, U. S. Patent 3,534,012 (1970); *CA* **74**, 42847q.
128. H. W. McCormick, *J. Polym. Sci.* **25**, 488 (1957); **41**, 327 (1959).
129. A. F. Sirianni, D. J. Worsfold, and S. Bywater, *Trans. Faraday Soc.* **55**, 2124 (1959).
130. D. J. Worsfold and S. Bywater, *J. Polym. Sci.* **26**, 299 (1957).
131. D. J. Worsfold and S. Bywater, *Can. J. Chem.* **36**, 1141 (1958).

132. F. Wenger, *Makromol. Chem.* **37**, 143 (1960).
133. F. Wenger, *J. Am. Chem. Soc.* **82**, 4281 (1960).
134. L. J. Fetters, L. A. Wall, and S. Strauss, umpublished data.
135. C. L. Lee, J. Smid, and M. Szwarc, *J. Phys. Chem.* **66**, 904 (1962).
136. A. Vranken, J. Smid, and M. Szwarc, *Trans. Faraday Soc.* **58**, 2036 (1962).
137. D. P. Wyman and I. H. Song, *Makromol. Chem.* **115**, 64 (1968).
138. H. G. Elias and V. S. Kamat, *Makromol. Chem.* **117**, 61 (1968).
139. Ruetgerswerke und Teerverwertung AG, British Patent 1,099,702 (1968); *CA* **68**, 50320d.
140. M. Tomoi and H. Kakiuchi, *Polym. J.* **5**, 193 (1973).
141. K. M. Hui and Y. K. Ong, *J. Polym. Sci., Polym. Chem. Ed.* **14**, 1311 (1976).
142. K. M. Hui and T. L. Ng, *J. Polym. Sci., Polym. Chem. Ed.* **7**, 3101 (1969).
143. J. M. Ginn and K. J. Ivin, *Makromol. Chem.* **139**, 47 (1970).
144. J. Leonard and S. L. Malhotra, *J. Polym. Sci., Polym. Chem. Ed.* **9**, 1983 (1971).
145. I. Mita and H. Okuyama, *J. Polym. Sci., Polym. Chem. Ed.* **9**, 3437 (1971).
146. G. B. Stampa, Ger. Offen. 1,943,266 (1970); *CA* **73**, 26057f.
147. Y. Okamoto, H. Yuki, and S. Murahashi, *Bull. Chem. Soc. Jpn.* **4**, 197 (1968).
148. M. Nakayama, H. Hirohara, K. Tayaka, and N. Ise, *J. Polym. Sci., Polym. Chem. Ed.* **8**, 3653 (1970).
149. H. Hirohara, M. Nakayama, R. Kawabata, and N. Ise, *J. Chem. Soc., Faraday Trans. 1* **68**, 51 (1972).
150. J. Geerts, M. van Beylan, and G. Smets, *J. Polym. Sci., Polym. Chem. Ed.* **7**, 2859 (1969).
151. R. W. H. Berry and R. J. Mazza, *Polymer* **14**, 172 (1973).
151a. M. E. Carter, J. L. Nash, Jr., J. W. Duerke, Jr., J. W. Schwietert, and G. B. Butler, *J. Polym. Sci., Polym. Chem. Ed.* **16**, 739 (1978).
152. W. Fowells, C. Schuerch, F. A. Borey, and F. P. Hood, *J. Am. Chem. Soc.* **89**, 1396 (1967).
152a. K. Hatada, Y. Umemura, M. Furomoto, S. Kokan, K. Ohta, and H. Yuki, *Makromol. Chem.* **178**, 1215 (1977).
153. J. Heller and C. B. Kingsley, *Makromol. Chem.* **78**, 47 (1964).
154. G. F. D'Alelio and R. C. Evers, *J. Polym. Sci., Polym. Chem. Ed.* **5**, 813, 999 (1967).
155. Y. Miura, M. Kinoshita, and M. Imoto, *Makromol. Chem.* **157**, 51 (1972).
156. G. R. Dever, F. E. Karasz, W. J. McKnight, and R. W. Lenz, *J. Polym. Sci., Polym. Chem. Ed.* **13**, 2151 (1975).
157. K. Matsuzaki, T. Uryu, and K. Ito, *Makromol. Chem.* **126**, 292 (1969).
158. T. Tsuruta, Y. Kawakami, and R. Tsushima, *Makromol. Chem.* **149**, 135 (1971).
159. D. S. Breslow and A. Kutner, *J. Polym. Sci., Part B* **9**, 129 (1971).
160. K. Chikanishi, T. Tsuruta, and J. Furukawa, *Makromol. Chem.* **80**, 158 (1964).
161. B. Wesslen and R. W. Lenz, *Macromolecules* **4**, 20 (1971).
162. M. Ynoezawa, S. Suzuki, H. Ito, and K. Ito, *Yuki Gosei Kagaku Kyokaishi* **27**, 280 (1969); *CA* **70**, 106918r.
163. C. J. Buck, Braz. Pedido PI 7,406,220 (1976); *CA* **85**, 143823m.
164. J. Junquera, N. Cardona, and J. E. Figueruelo, *Makromol. Chem.* **160**, 159 (1972).
165. G. F. D'Alelio and T. R. Hoffend, *J. Polym. Sci., Polym. Chem. Ed.* **5**, 323 (1967).
165a. H. Yuki, K. Hatada, K. Ohta, and Y. Okamoto, *J. Macromol. Sci.* **A9**, 983 (1975).
166. H. Yuki, K. Hatada, T. Niinomi, and Y. Kikuchi, *Polym. J.* **1**, 36 (1970).
167. T. E. Lipatova, G. S. Shapoval, N. P. Bazilevskaya, E. S. Shevchuk, and A. E. Nesterov, *Sint. Fiz.-Khim. Polim.* **11**, 3 (1973); *CA* **80**, 3825k.
168. T. E. Lipatova, G. S. Shapoval, N. P. Bazilevskaya, and E. S. Shevchuk, *Nuova Chim.* **49**, 87 (1973); *CA* **79**, 53843q.
169. T. E. Lipatova, G. S. Shapoval, N. P. Bazilevskaya, and E. S. Shevchuk, *Vysokomol. Soedin., Ser. A* **14**, 2610 (1972); *CA* **78**, 72918p.
170. T. E. Lipatova and N. P. Basilevskaya, *J. Macromol. Sci., Chem. A* **10**, 1257 (1976).

171. T. Jiri and P. Vlcek, *Chem. Prum.* **18**, 312 (1968); *CA* **69**, 27923q.
172. T. Shimomura, K. Ono, E. Tsuchida, and I. Shinohara, *Kogyu Kagaku Zasshi* **71**, 1070 (1968); *CA* **69**, 97251s.
173. Ceskoslovenska Akademie Ued., Netherlands Patent Appl. 6,609,329 (1967); *CA* **67**, 3325d.
174. N. A. Mukhitdinova, A. Nazarov, and M. A. Askarov, *Uzb. Khim. Zh.* **17**, 57 (1973); *CA* **79**, 53858y.
175. J. Trekoval, P. Vlcek, and D. Lim, *Collect. Czech. Chem. Commun.* **36**, 3032 (1971).
176. G. Beinert, G. Hild, and P. Rempp, *Makromol. Chem.* **175**, 2069 (1974).
177. H. Yuki, Y. Okamoto, Y. Shimada, K. Ohta, and K. Hatada, *Polymer* **17**, 618 (1976).
178. T. Saegusa, S. Kobayashi, and Y. Kimura, *Macromolecules* **8**, 950 (1975).
179. K. Yamaguchi, Y. Mitsuda, and M. Kinoshita, *J. Macromol. Sci., Chem. A* **10**, 975 (1976).
180. P. Vlcek and J. Trekoval, *Makromol. Chem.* **176**, 2595 (1975).
181. N. Kawabata and T. Tsuruta, *Makromol. Chem.* **86**, 231 (1965).
182. W. K. Busfield and J. M. Methuen, *Polymer* **14**, 137 (1973).
183. H. Gisser and H. E. Mertwoy, *Macromolecules* **7**, 431 (1974).
184. K. Chikanishi and T. Tsuruta, *Makromol. Chem.* **81**, 198 (1965).
185. K. Hatada, S. Kokan, T. Niinomi, K. Miyaji, and H. Yuki, *J. Polym. Sci., Polym. Chem. Ed.* **13**, 2117 (1975).
186. H. Yuki, K. Ohta, K. Uno, and S. Murahashi, *J. Polym. Sci., Polym. Chem. Ed.* **6**, 829 (1968).
187. A. Rembaum and M. Szwarc, *J. Polym. Sci.* **22**, 189 (1956).
188. A. V. Tobolsky and D. B. Hartley, *J. Am. Chem. Soc.* **84**, 1391 (1962).
189. F. Wenger, *Chem. Ind. (London)* p. 1094 (1959).
190. D. L. Glusker, E. Stiles, and B. Yoncoskie, *J. Polym. Sci.* **49**, 297 (1961).
191. D. L. Glusker, I. Lysoff, and E. Stiles, *J. Polym. Sci.* **49**, 315 (1961).
192. D. L. Glusker, R. A. Galluccio, and R. A. Evans, *J. Am. Chem. Soc.* **86**, 187 (1964).
193. A. A. Korotkov, S. P. Mitsengendler, and V. N. Krasulina, *J. Polym. Sci.* **53**, 117 (1961).
194. D. M. Wiles and S. Bywater, *Polymer* **3**, 175 (1962).
195. B. J. Cottam, D. M. Wiles, and S. Bywater, *Can. J. Chem.* **41**, 1905 (1963).
196. W. E. Goode, F. H. Owens, and W. L. Myers, *J. Polym. Sci.* **47**, 75 (1960).
197. T. Tsuruta, T. Makimoto, and H. Kanai, *J. Macromol. Chem.* **1**, 31 (1966).
198. D. M. Wiles and S. Bywater, *Chem. Ind. (London)* p. 1209 (1963).
199. D. M. Wiles and S. Bywater, *J. Phys. Chem.* **68**, 1983 (1964).
200. D. M. Wiles and S. Bywater, *Trans. Faraday Soc.* **61**, 150 (1965).
201. R. K. Graham, D. L. Dunkelberger, and E. S. Cohn, *J. Polym. Sci.* **42**, 501 (1960).
202. A. Roig, J. E. Figueruelo, and E. Llano, *J. Polym. Sci., Part B* **3**, 171 (1965).
203. P. Rempp, V. I. Volkov, J. Parrod, and C. Sadron, *Bull. Chem. Soc. Fr.* p. 919 (1960).
204. W. E. Goode, F. H. Owens, and R. C. Fettes, *J. Polym. Sci.* **42**, 367 (1960).
205. H. Schreiber, *Makromol. Chem.* **36**, 86 (1959).
206. W. E. Goode, F. H. Owens, R. P. Fellmann, W. H. Snyder, and J. E. Moore, *J. Polym. Sci.* **46**, 317 (1960).
207. R. G. Miller, B. Mills, P. A. Small, A. Turner-Jones, and P. G. M. Wood, *Chem. Ind. (London)* p. 1323 (1958).
208. B. S. Garrett, W. E. Goode, S. Gratch, J. F. Kincaid, C. L. Levesque, A. Spell, J. D. Stroupe, and W. H. Watanabe, *J. Am. Chem. Soc.* **81**, 1007 (1959).
209. M. Morton, R. A. Pett, and J. F. Fellers, *Prepr. IUPAC Macromol. Symp., 1966* Vol. 1, p. 69 (1966).
210. Y. Joh and Y. Kotake, *Macromolecules* **3**, 337 (1970).
211. Y. Kotake, Y. Joh, and F. Ide, *J. Polym. Sci., Part B* **8**, 101 (1970).
212. L. K. Golova, Yu. B. Amerik, and B. A. Krentsel, *Vysokomol. Soedin., Ser. B* **12**, 565 (1970); *CA* **73**, 992412.

213. Yu. B. Amerik, W. F. Reynolds, and J. E. Guillet, *J. Polym. Sci., Polym. Chem. Ed.* **9**, 531 (1971).
214. B. Francois, L. Christman, C. Claude, and J. Parrod, *Kinet. Mech. Polyreactions, Int. Symp. Macromol. Chem., Prepr., 1969* Vol. 2, p. 21 (1969); *CA* **75**, 77369h.
215. A. Yamada, M. Yanagita, and M. Hirose, *Kogyo Kagaku Zasshi* **74**, 1185 (1971).
216. P. Vlcek and J. Trekoval, *Collect. Czech. Chem. Commun.* **37**, 1918 (1972); *CA* **77**, 127113j.
217. J. Trekoval and P. Kratochvil, *J. Polym. Sci., Polym. Chem. Ed.* **10**, 1391 (1972).
218. P. Ederle, G. Friedmann, and M. Brini, *Bull. Soc. Chim. Fr.* **10**, 3992 (1972).
219. M. Morton, E. E. Bostick, R. A. Livigni, and L. J. Fetters, *J. Polym. Sci., Part A* **1**, 1735 (1963).
220. A. A. Korotkov and V. N. Krasulina, *Vysokomol. Soedin. Ser. A* **10**, 1570 (1968); *Polym. Sci. USSR (Engl. Transl.)* **10**, 1818 (1968).
221. T. Hirahara, T. Nakano, and Y. Minoura, *J. Polym. Sci., Polym. Chem. Ed.* **6**, 485 (1968).
222. V. Warzelhan and G. V. Schulz, *Makromol. Chem.* **177**, 2185 (1976).
223. S. Fujishige, *Makromol. Chem.* **177**, 375 (1976).
224. T. Miyamoto, S. Tomoshige, and H. Inagaki, *Makromol. Chem.* **176**, 3035 (1975).
225. G. Löhr, A. H. E. Müller, V. Warzelhan, and G. V. Schulz, *Makromol. Chem.* **175**, 497 (1974).
226. J. P. Pascault, F. Chastrette, and Q. T. Pham, *Eur. Polym. J.* **12**, 273 (1976).
227. I. B. Rashov, I. M. Panayotov, and P. H. Li, *Eur. Polym. J.* **11**, 361 (1975).
228. I. Mita, Y. Watabe, T. Akatsu, and H. Kamble, *Polym. J.* **4**, 271 (1973).
229. J. Kawak, Q. T. Pham, C. Pillot, and J. Pascault, *Eur. Polym. J.* **10**, 997 (1974).
230. S. N. Lewis and R. A. Haggard, Ger. Offen. 2,262,588 (1973); *CA* **80**, 71358k.
231. A. C. Angood, S. A. Hurley, and P. J. T. Tait, *J. Polym. Sci., Polym. Chem. Ed.* **11**, 2777 (1973).
232. G. Löhr and G. V. Schulz, *Eur. Polym. J.* **10**, 121 (1974).
233. K. A. Allen, B. G. Gowenlock, and W. E. Lindsell, *J. Polym. Sci., Polym. Chem. Ed.* **12**, 1131 (1974).
234. I. M. Panayotov and I. B. Rashkov, *J. Polym. Sci., Polym. Chem. Ed.* **11**, 2615 (1973).
235. T. Hirahara and Y. Minoura, *J. Polym. Sci., Polym. Chem. Ed.* **12**, 1395 (1974).
236. L. Lochmann, M. Rodova, and J. Trekoval, *J. Polym. Sci., Polym. Chem. Ed.* **12**, 2091 (1974).
237. L. Lochmann, M. Rodova, J. Peteranek, and D. Lim, *J. Polym. Sci., Polym. Chem. Ed.* **12**, 2295 (1974).
238. J. P. Pascault, J. Kawak, J. Gole, and Q. T. Pham, *Eur. Polym. J.* **10**, 1107 (1974).
239. G. Löhr and G. V. Schulz, *Makromol. Chem.* **172**, 137 (1973).
240. C. E. H. Bawn, A. Ledwith, and N. R. McFarlane, *Polymer* **8**, 484 (1967).
241. G. M. Guzman and A. Bello, *Makromol. Chem.* **107**, 46 (1967).
242. A. Roig, J. E. Figueruelo, and J. M. Gutierrez, *An. Real. Soc. Espan. Fis. Quim., Ser. B* **62**, 1081 (1966); *CA* **66**, 116017c.
243. B. D. Coleman, T. G. Fox, and M. Reinmoeller, *J. Polym. Sci. B* **4**, 1029 (1966).
244. A. Roig and J. E. Figueruelo, *An. R. Soc. Esp. Fis. Quim., Ser. B* **62**, 1073 (1966).
245. T. Yamaguchi and Y. Hata, *Kobunshi Kagaku* **24**, 203 (1967).
246. T. E. Lipatova, G. S. Shapoval, E. S. Shevchuk, and N. P. Bazilevskaya, *J. Macromol. Sci., Chem. A* **5**, 345 (1971).
247. H. Kaemmerer, H. Gueniffey, and C. Pinazzi, *C. R. Hebd. Seances Acad. Sci., Ser. C* **272**, 1714 (1971); *CA* **75**, 130151j.
248. A. A. Korotkov, Z. A. Azimov, and S. P. Mitsengendler, *Vysokomol. Soedin.* **7**, 1326 (1965); *Polym. Sci. USSR (Engl. Transl.)* **7**, 1740 (1966).
249. G. Natta, M. Farina, and M. Donati, *Makromol. Chem.* **43**, 251 (1961).
250. O. H. Griffith, J. F. W. Keana, S. Rottschaefer, and T. A. Warlick, *J. Am. Chem. Soc.* **89**, 5072 (1967).

251. H. Gueniffey, E. Klesper, and H. Kammerer, *Makromol. Chem.* **162**, 199 (1972).
252. H. Yuki, K. Hatada, Y. Kikuchi, and T. Niinomi, *J. Polym. Sci., Part B* **6**, 753 (1968).
253. O. Mamontova, A. Abkin, and S. Medvedev, *Acta Physicochim. URSS* **12**, 269 (1940).
254. A. Abkin and S. Medvedev, *Trans. Faraday Soc.* **32**, 286 (1936); *Zh. Fiz. Khim.* **13**, 705 (1939).
255. W. Schlenk and E. Bergmann, *Justus Liebigs Ann. Chem.* **463**, 1 (1928).
256. K. Ziegler and K. Bahr, *Ber. Dtsch. Chem. Ges.* **61**, 263 (1928).
257. K. Ziegler, F. Dersch, and H. Wollthan, *Justus Liebigs Ann. Chem.* **473**, 36 (1929).
258. K. Ziegler and O. Schafer, *Justus Liebigs Ann. Chem.* **479**, 150 (1930).
259. K. Ziegler, F. Dersch, and H. Wollthan, *Justus Liebigs Ann. Chem.* **511**, 13 (1934).
260. K. Ziegler and L. Jacob, *Justus Liebigs Ann. Chem.* **511**, 45 (1934).
261. K. Ziegler, L. Jacob, H. Wollthan, and A. Wenz, *Justus Liebigs Ann. Chem.* **511**, 64 (1934).
262. W. Schlenk and E. Bergmann, *Justus Liebigs Ann. Chem.* **479**, 42, 58, 78 (1930).
263. K. Ziegler, *Angew. Chem.* **49**, 499 (1936).
264. R. E. Robertson and L. Marion, *Can. J. Res., Sect. B* **26**, 657 (1948).
265. K. Ziegler and L. Jacob, *Justus Liebigs Ann. Chem.* **511**, 52 (1934).
266. R. V. Basova, A. A. Arest-Yakubovich, D. A. Solovykh, N. V. Desytova, A. R. Gantmakher, and S. S. Medvedev, *Dokl. Akad. Nauk SSSR* **149**, 1067 (1963).
267. H. L. Hsieh, *J. Polym. Sci., Part A* **3**, 163, 181, 191 (1965).
268. I. Kuntz and A. Gerber, *J. Polym. Sci.* **42**, 299 (1960).
269. A. F. Johnson and D. J. Worsfold, *J. Polym. Sci., Part A* **3**, 449 (1965).
270. Y. Watabe, M. Ishii, Y. Iseda, K. Komatsu, N. Ohshima, and Y. Nakata, Ger. Offen. 2,640,645 (1977); *CA* **86**, 156805c.
271. I. G. Hargis and R. A. Livigni, U. S. Patent 3,966,638 (1976); *CA* **85**, 94953t.
272. J. Q. Wood, U. S. Patent 3,382,224 (1968); *CA* **69**, 11236g.
273. H. S. Makowski, M. Lynn, and A. N. Bogard, *J. Macromol. Sci., Chem. A* **2**, 665 (1968).
274. H. S. Makowski and M. Lynn, *J. Macromol. Sci.* **1**, 443 (1966); *J. Macromol. Sci., Chem. A* **2**, 683 (1968).
275. Esso Research and Engineering Co. French Patent 1,492,154 (1967); *CA* **68**, 106145v.
276. U. Gohlke, T. Win, and G. Reinisch, *Faserforsch. Textiltech.* **25**, 191 (1974); *CA* **81**, 121177u.
277. V. N. Zgonnik, K. K. Kalnin'sh, Ye. Yu. Shadrina, N. V. Smirnova, and N. I. Nikolayev, *Vysokomol. Soedin., Ser. A* **16**, 1867 (1974); *Polym. Sci. USSR* (*Engl. Transl.*) *A* **16**, 2163 (1974).
278. V. N. Zgonnik, K. K. Kalnin'sh, Kh. B. Tsvetanov, and N. I. Nikolayev, *Vysokomol. Soedin., Ser. A* **15**, 900 (1973); *Polym. Sci. USSR* (*Engl. Transl.*) **15**, 1013 (1973).
279. J. Sledz, F. Schue, B. Kampf, and S. Libs, *Eur. Polym. J.* **10**, 1207 (1974).
280. V. P. Shatalov, I. Yu. Kirchevskaya, A. R. Samotsvetov, and N. P. Proskurina, *Vysokomol. Soedin., Ser. A* **15**, 2042 (1973); *Polym. Sci. USSR* (*Engl. Transl.*) **15**, 2309 (1973).
281. B. I. Nakhmanovich, A. M. Sakharov, and A. A. Arest-Yakubovich, *Dokl. Akad. Nauk SSSR* **212**, 1379 (1973); *CA* **80**, 133903v.
282. B. I. Nakhmanovich, L. N. Moskalenko, and A. A. Arest-Yakubovich, *Dokl. Akad. Nauk SSSR* **215**, 137 (1974).
283. F. E. Naylor, H. L. Hsieh, and J. C. Randall, *Macromolecules* **3**, 486 (1970).
284. J. C. Randall and R. S. Silas, *Macromolecules* **3**, 491 (1970).
285. J. C. Randall, F. E. Naylor, and H. L. Hsieh, *Macromolecules* **3**, 497 (1970).
286. S. Bywater, D. J. Worsfold, and G. Hollingsworth, *Macromolecules* **5**, 389 (1972).
287. M. Morton, R. D. Sanderson, and R. Sakata, *J. Polym. Sci., Part B* **9**, 61 (1971); *Macromolecules* **6**, 181 (1973).
288. W. H. Glaze, J. E. Hanicak, M. L. Moore, and J. Chaudhuri, *J. Organomet. Chem.* **44**, 39 (1972).
289. Michelin et Cie., German Patent 1,620,987 (1970); *CA* **74**, 43302v.

290. Y. Hoshirro, Y. Iwaki, K. Matsukawa, A. Nakada, F. Sato, and K. Uno, Ger. Offen. 2,011,561 (1970); *CA* **74**, 4014h.
291. L. N. Moskalenko and A. A. Arest-Yakubovich, *Dokl. Akad. Nauk SSSR* **195**, 1370 (1970).
292. I. A. Livshits and Y. B. Podolnyi, *Vysokomol. Soedin., Ser. A* **12**, 2655 (1970); *Polym. Sci. USSR (Engl. Transl.)* **12**, 3008 (1970).
293. E. Okutani, F. Ito, A. Sakaguchi, K. Komatsu, and H. Yasuhaga, Japanese Patent 70/20,299 (1970); *CA* **74**, 64635w.
294. C. A. Uraneck, *J. Polym. Sci., Polym. Chem. Ed.* **9**, 2273 (1971).
295. T. A. Antkowiak, A. E. Oberster, A. F. Halasa, and D. P. Tate, *J. Polym. Sci., Polym. Chem. Ed.* **10**, 1319 (1972).
296. W. J. Kern, J. N. Anderson, H. E. Adams, T. C. Bouton, and T. W. Bethea, *Polym. Prepr., Am. Chem. Soc., Div. Polym. Chem.* **12**(1), 449 (1971).
297. R. V. Basova, N. V. Kozlova, and A. R. Gantmakher, *Vysokomol. Soedin., Ser. A* **15**, 1234 (1973); *Polym. Sci. USSR (Engl. Transl.)* **15**, 1385 (1973).
298. H. L. Hsieh and C. F. Wofford, *J. Polym. Sci., Polym. Chem. Ed.* **7**, 449 (1969).
299. A. Garton and S. Bywater, *Macromolecules* **8**, 694 (1975).
300. A. Garton and S. Bywater, *Macromolecules* **8**, 697 (1975).
301. N. Sommer and K. H. Nordsiek, Ger. Offen. 2,158,574; 2,158,575 (1973); *CA* **79**, 93216u, 54635k.
302. T. Narita, Y. Kunitake, and T. Tsuruta, *Makromol. Chem.* **176**, 3371 (1975).
303. F. C. Foster, U. S. Patent 3,317,918 (1967); *CA* **67**, 22713v.
304. T. C. Chen and A. F. Halasa, *J. Polym. Sci., Polym. Chem. Ed.* **14**, 573 (1976).
305. T. C. Chen and A. F. Halasa, *J. Polym. Sci., Polym. Chem. Ed.* **14**, 583 (1976).
306. B. I. Nakhmanovich, R. V. Basova, and A. A. Arest-Yakubovich, *J. Macromol. Sci., Chem. A* **9**, 575 (1975).
307. H. L. Hsieh, *J. Polym. Sci., Polym. Chem. Ed.* **14**, 379 (1976).
308. D. N. Schulz, A. F. Halasa, and A. E. Oberster, *J. Polym. Sci., Polym. Chem. Ed.* **12**, 153 (1974).
309. Firestone Tire and Rubber Co., Netherlands Patent Appl. 6,600,822 (1966); *CA* **66**, 19552.
310. G. R. Kahle, U. S. Patent 3,296,150 (1967); *CA* **66**, 46766v.
311. Chemische Werke Huels, AG, French Patent 1,499,383 (1966); *CA* **66**, 56384h.
312. W. T. House, U. S. Patent 3,288,872 (1966); *CA* **66**, 19042z.
313. I. G. Hargis and R. A. Livigni, U. S. Patent 4,012,336 (1977); *CA* **86**, 156238v.
314. M. H. Martin, *J. Polym. Sci., Polym. Chem. Ed.* **10**, 455 (1972).
315. B. L. Yerusalimskii, I. G. Krasnosel'skaya, and V. V. Mazurek, *Vysokomol. Soedin.* **6**, 1294 (1964).
316. I. G. Krasnosel'skaya and B. L. Yerusalimskii, *Vysokomol. Soedin.* **6**, 1637 (1964); *Polym. Sci. USSR (Engl. Transl.)* **6**, 1814 (1964).
317. I. G. Krasnosel'skaya and B. L. Yerusalimskii, *Vysokomol. Soedin., Ser. A* **9**, 851 (1967); *Polym. Sci. USSR (Engl. Transl.)* **6**, 952 (1967).
318. I. G. Krasnosel'skaya, B. L. Yerusalimskii, and V. V. Mazurek, *Vysokomol. Soedin., Ser A* **9**, 2130 (1967); *Polym. Sci. USSR (Engl. Transl.)* **9**, 2406 (1967).
319. R. Worley and R. N. Young, *Eur. Polym. J.* **8**, 1355 (1972).
320. U. Giannini, M. Cambini, and A. Cassata, *Makromol. Chem.* **61**, 246 (1963).
321. B. E. Erofeev, S. V. Naumova, and I. V. Kulevskaya, USSR Patent 191,797 (1967); *CA* **68**, 30612x.
322. A. R. Khalilpour, G. Jenner, and A. Deluzavche, *Bull. Soc. Chim. Fr.* p. 583 (1976); *CA* **85**, 33510z.
323. H. Yuki, Y. Okamoto, and H. Tokano, *Polym. J.* **2**, 663 (1971).
324. N. Roy and J. Prud'homme, *Macromolecules* **8**, 78 (1975).
325. R. Methin, B. Kaempf, and C. Tanielian, *Eur. Polym. J.* **13**, 493 (1977).

326. G. Jenner and A. Khalilpour, *Eur. Polym. J.* **12**, 105 (1976).
327. I. A. Livshits, V. I. Stepanova, K. V. Nelson, and A. P. Berezkina, *Vysokomol. Soedin., Ser. A* **9**, 2521 (1967).
327a. M. Morton, R. D. Sanderson, R. Sakata, and L. A. Falvo, *Macromolecules* **6**, 186 (1973).
328. N. G. Gaylord and B. Matyska, *J. Polym. Sci., Part B* **5**, 131 (1967).
329. V. L. Bell, *J. Polym. Sci., Part A* **2**, 5291 (1964).
330. R. P. Rhodes and D. A. Guthrie, U. S. Patent 3,346,664 (1967); *CA* **68**, 30890m.
331. C. Harries, U. S. Patent 1,058,056 (1913); *Justus Liebigs Ann. Chem.* **383**, 213 (1911).
332. F. Mathews and E. Strange, British Patent 24,790 (1910).
333. F. C. Foster and J. R. Binder, *Adv. Chem. Ser.* **19**, 26 (1957).
334. F. W. Stavely *et al.*, *Ind. Eng. Chem.* **48**, 778 (1956).
335. V. Sinn and J. Minoux, *C. R. Hebd. Seances Acad. Sci.* **251**, 2020 (1960).
336. J. Minoux, *Rev. Gen. Caoutch.* **39**, 779 (1962).
337. R. P. Foss, H. W. Jacobson, and W. H. Sharkey, *Macromolecules* **10**, 287 (1977).
338. G. Jenner, J. Hitzke, and M. Millet, *Bull. Soc. Chim. Fr.* p. 1183 (1970).
339. A. B. Moustafa, *Indian J. Technol.* **8**, 290 (1970); *CA* **74**, 42706t.
340. G. Jenner, *J. Macromol. Sci., Chem.* **A9**, 83 (1975).
341. L. M. Lanovskaya, Z. P. Chinkova, N. P. Slepova, L. A. Galata, and L. S. Kofman, *Vysokomol. Soedin., Ser. A* **17**, 896 (1975); *Polym. Sci. USSR (Engl. Transl.)* **17**, 1032 (1976).
342. J. E. L. Roovers and S. Bywater, *Macromolecules* **1**, 328 (1968).
343. J. W. A. Averink and P. J. Kok, British Patent 1,111,978 (1968); *CA* **69**, 3607p.
344. A. Guyot and J. Vialle, *J. Polym. Sci., Part B* **6**, 403 (1968).
345. Y. Tanaka and S. Otsuka, *Kobunshi Kagaku* **25**, 177, 355, 364 (1968).
346. M. Morton, L. J. Fetters, R. A. Pett, and J. F. Meier, *Macromolecules* **3**, 327 (1970).
347. M. Morton, R. A. Pett, and L. J. Fetters, *Macromolecules* **3**, 333 (1970).
348. D. J. Worsfold and S. Bywater, *Macromolecules* **5**, 393 (1972).
349. F. Schue, D. J. Worsfold, and S. Bywater, *J. Polym. Sci., Part B* **7**, 821 (1969).
350. B. Morese-Seguela, M. St-Jacques, J. M. Renaud, and J. Prud'homme, *Macromolecules* **10**, 431 (1977).
351. A. Davidjan, N. Nikolaew, V. Sgonnik, B. Belenkii, V. Nesterow, and B. Erussalimsky, *Makromol. Chem.* **177**, 2469 (1976).
352. A. Garton, R. P. Chaplin, and S. Bywater, *Eur. Polym. J.* **12**, 697 (1976).
353. A. C. Angood, S. A. Hurley, and P. J. T. Tait, *J. Polym. Sci., Polym. Chem. Ed.* **11**, 2777 (1973); **13**, 2437 (1975).
354. R. Salle and Q. T. Pham, *J. Polym. Sci., Polym. Chem. Ed.* **15**, 1799 (1977).
355. A. Essel, R. Salle, J. Gole, and Q. T. Pham, *J. Polym. Sci., Polym. Chem. Ed.* **13**, 1847, 1855, 1869 (1975).
356. A. Essel, Q. T. Pham, and J. Gole, *J. Polym. Sci., Polym. Chem. Ed.* **11**, 1851 (1973).
357. J. M. Alvarino, A. Bello, and G. M. Guzman, *An. Quim.* **70**, 565 (1974); *CA* **82**, 4645e.
358. F. Schue, D. J. Worsfold, and S. Bywater, *Macromolecules* **3**, 509 (1970).
359. A. Guyot and J. Vialle, *J. Macromol. Sci., Chem.* **A4**, 79, 107 (1970).
360. C. Stein and J. Gole, *Bull. Soc. Chim. Fr.* p. 3175 (1966).
360a. M. Morton and L. A. Falvo, *Macromolecules* **6**, 190 (1973).
361. I. A. Arbuzova, V. N. Efremova, A. G. Eliseeva, N. V. Mikhailova, V. N. Nikitin, A. V. Sidorovich, N. A. Klushin, and E. V. Kuvshinskii, *Vysokomol. Soedin., Ser. A* **12**, 697 (1970).
362. I. A. Livshits and T. T. Denisova, *Vysokomol. Soedin., Ser. B* **14**, 317 (1972); *CA* **77**, 48844v.
363. K.-F. Elgert and W. Ritter, *Makromol. Chem.* **177**, 2021 (1976).
364. A. Ulrich, A. Deluzarehe, A. Maillard, F. Schue, and C. Tanielian, *Bull. Soc. Chim. Fr.* p. 2460 (1972).

365. I. A. Livshits, T. T. Denisova, and V. E. Stepanova, *Vysokomol. Soedin., Ser. A* **14**, 928 (1972).
366. D. Cuzin, Y. Chauvin, and G. Lefebvre, *Eur. Polym. J.* **3**, 581 (1967).
367. I. A. Livshits and V. I. Stepanova, *Vysokomol. Soedin.* **7**, 181 (1965); *Polym. Sci. USSR (Engl. Transl.)* **7**, 197 (1965).
368. B. Kaempf, A. Maillard, F. Schue, J. Sledz, J. Sommer, and C. Tanielian, *Bull. Soc. Chim. Fr.* p. 1153 (1972); *CA* **77**, 48874e.
369. R. J. Ambrose and W. L. Hergenrother, *Macromolecules* **5**, 275 (1972).
370. P. de Radzitzki and G. Smets, *Bull. Soc. Chim. Belg.* **62**, 320 (1953).
371. R. Asami and A. Shoji, *Int. Symp. Macromol. Chem., 1966* (1966).
372. R. Radovanovic-Kiprijanova and D. Fles, *J. Polym. Sci., Polym. Chem. Ed.* **13**, 1141 (1975).
373. W. E. Hanford, J. R. Roland, and H. S. Young, U. S. Patent 2,377,779 (1945).
374. A. Langer, *Trans. N. Y. Acad. Sci.* [2] **27**, 741 (1965).
375. A. Langer, *Polym. Prepr., Am. Chem. Soc., Div. Polym. Chem.* **7**, 132 (1966).
376. K. Ziegler and H. G. Gellert, *Justus Liebigs Ann. Chem.* **567**, 179, 185, 195 (1950).
377. P. D. Bartlett, S. Friedman, and M. Stiles, *J. Am. Chem. Soc.* **75**, 1771 (1954).
378. G. G. Eberhardt and W. R. Davis, *J. Polym. Sci., Part A*, **3**, 3753 (1965).
379. H. Magnin, F. Rodriguez, M. Abadie, and F. Schue, *J. Polym. Sci., Polym. Chem. Ed.* **15**, 875, 897, 901 (1977).
380. J. N. Hay, D. S. Harris, and D. M. Wiles, *Polymer* **17**, 613 (1976).
381. J. N. Hay, J. F. McCabe, and J. C. Robb, *J. Chem. Soc., Faraday Trans. I* **68**, 1227 (1972).
382. G. G. Eberhardt and W. A. Butte, *J. Org. Chem.* **29**, 2928 (1964).
383. F. Rodriguez, M. Abadie, and F. Schue, *Eur. Polym. J.* **12**, 17, 23 (1976).
384. F. Rodriguez, M. Abadie, and F. Schue, *J. Polym. Sci., Polym. Chem. Ed.* **14**, 773 (1976).
385. A. Rembaum and J. Moacanin, *J. Polym. Sci., Part B* **2**, 979 (1964).
386. V. M. Story and G. Canty, *J. Res. Natl. Bur. Stand. Sect. A* **68**, 165 (1964).
387. K. Butler, P. R. Thomas, and G. J. Tyler, *J. Polym. Sci.* **48**, 357 (1960).
388. H. Kobayashi, K. Yamaguchi, C. Ohizumi, and J. Masamota, Japanese Patent 71/21,739 (1971); *CA* **76**, 15174c.
389. G. M. Chetyrkina, I. L. Artamonova, K. Kalnins, and A. V. Sidorovich, *Vysokomol. Soedin., Ser. B* **10**, 150 (1968); *CA* **69**, 3197e.
390. H. Nakayama, T. Higashimura, and S. Okamura, *Kobunshi Kagaku* **23**, 439 (1966).
391. H. Nakayama, Y. Yamasawa, T. Higashimura, and S. Okamura, *Kobunshi Kagaku* **24**, 290 (1967).
392. S. Dabi and A. Zilkha, *Eur. Polym. J.* **13**, 545 (1977).
393. G. Camino, S. L. Lim, and L. Trossarelli, *Eur. Polym. J.* **13**, 479 (1977).
394. L. Trossarelli, M. Guaita, and G. Camino, *Makromol. Chem.* **105**, 285 (1967).
395. S. Kadowaki, M. Kameya, H. Kitakami, and T. Nakamura, Japanese Patent 76/40,114 (1976); *CA* **86**, 140681y.
396. L. W. Bush and D. S. Breslow, *Macromolecules* **1**, 189 (1968).
397. T. Otsu, B. Yamada, M. Itahashi, and T. Mori, *J. Polym. Sci., Polym. Chem. Ed.* **14**, 1347 (1976).
398. A. Zilkha and A. Ottolenghi, *J. Polym. Sci.*, **56**, 539 (1962); *J. Polym. Sci., Part A* **1**, 687 (1963).
399. A. Zilkha and Y. Avny, *J. Polym. Sci., Part A* **1**, 549 (1963).
400. M. L. Miller, *J. Polym. Sci.*, **56**, 203 (1962).
401. A. Ottolenghi, S. Barzakay, and A. Zilkha, *J. Polym. Sci., Part A* **1**, 3643 (1963).
402. J. Furukawa, T. Saegusa, and H. Mise, *Makromol. Chem.* **38**, 224 (1960).
403. A. Zilkha and B. A. Feit, *J. Appl. Polym. Sci.* **5**, 252 (1961); **7**, 287 (1963).
404. A. Tsukamoto, *J. Polym. Sci., Part A* **3**, 2767 (1965).
405. P. Rempp, *Polym. Prepr., Am. Chem. Soc., Div. Polym. Chem.* **7**, 141 (1966).

406. R. B. Cundall, J. Driver, and D. D. Eley, *Proc. Chem. Soc., London* p. 170 (1958).
407. B. L. Erussalimski and I. G. Krasnosel'skaya, *Makromol. Chem.* **123**, 80 (1969).
408. B. L. Erussalimski, I. V. Kulevskaya, and V. V. Masurek, *J. Polym. Sci., Part C* **16**, 1355 (1967).
409. S. E. Bresler, B. L. Erussalimski, and I. V. Kulevskaya, *J. Polym. Sci., Polym. Chem. Ed.* **6**, 2795 (1968).
410. T. Hirahara and Y. Minoura, *J. Polym. Sci., Polym. Chem. Ed.* **8**, 3391 (1970).
411. A. V. Novoselova, B. L. Erussalimski, V. N. Krasulina, and E. V. Zashcherinskii, *Vysokomol. Soedin., Ser. A* **13**, 87 (1971); *CA* **74**, 100418p.
412. B. L. Erussalimski, I. G. Krasnosel'skaya, V. N. Krasulina, A. V. Novoselova, and E. V. Zashcherinskii, *Eur. Polym. J.* **6**, 1391 (1970).
413. K. Yamaguchi, O. Sonoda, and Y. Minoura, *J. Polym. Sci., Polym. Chem. Ed.* **10**, 63 (1972).
414. C. D. Eisenbach, V. Jaacks, H. Schnecko, and W. Kern, *Makromol. Chem.* **175**, 1329 (1974).
415. V. Jaacks, C. D. Eisenbach, and W. Kern, *Makromol. Chem.* **161**, 139 (1972).
416. A. V. Novoselova, P. P. Nefedov, S. Ya. Frenkel, M. A. Kuznetsova, and B. L. Erussalimski, *Vysokomol. Soedin., Ser. B* **16**, 751 (1974).
417. N. Kawabata and T. Tsuruta, *Makromol. Chem.* **98**, 262 (1966).
418. A. V. Novoselova, B. L. Erussalimski, V. N. Krasulina, and Ye. V. Zashcherinskii, *Vysokomol. Soedin., Ser. A* **13**, 87 (1971); *Polym. Sci. USSR* (*Engl. Transl.*) **13**, 99 (1971).
419. C. D. Eisenbach, G. Franzmann, V. Jaacks, H. Schnecko, and W. Kern, *Makromol. Chem.* **175**, 1789 (1974).
420. H. Balard and J. Meybeck, *Eur. Polym. J.* **13**, 611 (1977).
421. T. Ogawa and P. Quintana, *J. Polym. Sci., Polym. Chem. Ed.* **13**, 2517 (1975).
422. T. Hirahara, H. Inoue, and Y. Minoura, *Eur. Polym. J.* **10**, 109 (1974).
423. K. Tsvetanov, V. N. Zgonnik, B. L. Erussalimski, and I. Panaiotov, *Adv. Ionic Polym., Proc. Int. Symp., 1972*, p. 65 (1975).
424. M. Ragazzini, A. Vandi, and F. Campadelli, *Eur. Polym. J.* **6**, 1331 (1970).
425. I. P. Gozman, L. K. Gazizova, Yu. A. Levin, T. A. Yagfarova, V. A. Byler, and B. E. Ivanov, *Sb. Nek. Probl. Org. Fiz. Khim., Mater. Nauchn. Sess., Inst. Org. Fiz. Khim., Akad. Nauk SSSR, 1972*, p. 85 (1972); *CA* **78**, 84841w.
426. J. Kovacs and C. S. Marvel, *J. Polym. Sci., Polym. Chem. Ed.* **5**, 563 (1967).
427. M. L. Miller and J. Skogman, *J. Polym. Sci., Part A* **2**, 4551 (1964).
428. T. Kitano, T. Fujimoto, and M. Nagasawa, *Macromolecules* **7**, 719 (1974).
429. J. H. Koral, *Polym. Prepr., Am. Chem. Soc., Div. Polym. Chem.* **3**, No. 2, 293 (1962).
430. T. Makimoto, T. Miyazako, and T. Tsuruta, *Kogyo Kagaku Zasshi* **70**, 199 (1967).
431. Y. Kobuke, J. Furukawa, and T. Fueno, *J. Polym. Sci., Polym. Chem. Ed.* **5**, 2701 (1967).
432. J. Manassen and J. Wallach, *J. Am. Chem. Soc.* **87**, 2671 (1965).
433. J. Wallach and J. Manassen, *J. Polym. Sci., Polym. Chem. Ed.* **7**, 1983 (1969).
434. L. E. Coleman, U. S. Patent 3,458,478 (1969); *CA* **71**, 71238k.
435. M. Benes, J. Peska, and O. Wichterle, *J. Polym. Sci., Part C* **4**, 1377 (1964).
436. V. Jaacks and G. Franzmann, *Makromol. Chem.* **143**, 283 (1971).
437. N. D. Scott, U. S. Patent 2,146,477 (1939).
438. N. D. Scott and J. F. Walker, *Ind. Eng. Chem.* **32**, 312 (1940).
439. L. A. Wall, L. J. Fetters, and S. Strauss, *J. Polym. Sci.* (to be published).
440. G. Champetier, M. Fontanille, A. Korn, and P. Sigwalt, *J. Polym. Sci.* **58**, 911 (1962).
441. S. F. Naumova, B. V. Erofeev, T. P. Maksimova, and Q. D. Yurina, USSR Patent 201,654 (1967); *CA* **68**, 78738v.
442. R. M. Nasirova, A. N. Vinogradov, E. A. Mushina, M. V. Shishkina, B. A. Krentsel, and V. A. Belyaev, *Vysokomol. Soedin., Ser. B* **14**, 857 (1972); *CA* **78**, 124933m.
443. A. Bachrach, Y. Tsur, and A. Zilkha, *Eur. Polym. J.* **12**, 563 (1976).

444. V. M. Misin, P. P. Kisilitsa, N. I. Bolondaeva, and M. I. Cherkashin, *Vysokomol. Soedin., Ser. A* **18**, 1726 (1976); *CA* **85**, 143576h.
445. S. Nakahama, A. Hirao, Y. Ohira, and N. Yamazaki, *J. Macromol. Sci., Chem. A* **9**, 563 (1975).
446. H. Höcker and G. Lattermann, *Makromol. Chem.* **158**, 191 (1972).
447. H. R. Allcock and R. L. Kugel, *J. Polym. Sci., Part A* **1**, 3627 (1963).
448. W. W. Limburg, J. F. Yonus, D. J. Williams, A. O. Goedde, and J. M. Pearson, *J. Polym. Sci., Polym. Chem. Ed.* **13**, 1133 (1975).
449. V. V. Korshak, L. V. Dzashi, and S. L. Sosin, *Nuova Chim.* **49**, 31 (1973).
450. K. Yagi, T. Miyazaki, H. Okitsu, F. Toda, and Y. Iwakura, *J. Polym. Sci., Polym. Chem. Ed.* **10**, 1149 (1972).
451. C. L. Schilling, Jr. and J. E. Mulvaney, *Macromolecules* **1**, 445 (1968).
452. C. H. Bamford, G. C. Eastmond, and Y. Imanishi, *Polymer* **8**, 651 (1967).
453. R. C. P. Cubbon, *Polymer* **6**, 419 (1965).
454. C. G. Overberger, W. M. Pearce, and N. Mayes, *J. Polym. Sci.* **31**, 217 (1958); **34**, 109 (1959).
455. R. G. Beaman, *J. Am. Chem. Soc.* **70**, 3115 (1948).
456. H. Sobue, T. Uryu, K. Matsuzaki, and Y. Tabata, *J. Polym. Sci., Part B* **1**, 409 (1963); *Part A* **2**, 3333 (1964).
457. F. Ranogajec, M. A. Markevich, E. V. Kochetov, and N. S. Enikolopyan, *Dokl. Akad. Nauk SSSR* **200**, 634 (1971); *CA* **76**, 60155h.
458. C. B. Tsvetanov and I. Panayotov, *Eur. Polym. J.* **11**, 209 (1975).
459. I. M. Panayotov, C. B. Tsvetanov, and A. C. Alexandrov, *Eur. Polym. J.* **11**, 875 (1975).
460. F. Ranogajec, E. V. Kochetov, M. A. Markevich, N. S. Enikolopyan, and I. Dvornik, *Croat. Chem. Acta* **46**, 83 (1974).
461. M. A. Markevich, E. V. Kochetov, F. Ranogajec, and N. S. Enikolopyan, *Adv. Ionic Polym., Proc. Int. Symp., 1972* p. 89 (1975).
462. V. E. Kaiser and R. C. Schulz, *Makromol. Chem.* **81**, 273 (1959).
463. M. Galin and J. C. Galin, *Eur. Polym. J.* **9**, 1149 (1973).
464. National Distillers and Chemical Corp. Belgian Patent 667,490 (1965); *CA* **65**, 9141c.
465. H. Wieland and E. Sakellarios, *Ber. Dtsch. Chem. Ges.* **52**, 898 (1919).
466. R. C. Sovish and W. Boettcher, *J. Polym. Sci. A* **2**, 5247 (1964).
467. A. T. Blomquist, W. J. Tapp, and J. R. Johnson, *J. Am. Chem. Soc.* **67**, 1519 (1945).
468. V. P. Alaniya, E. G. Gorlov, L. V. Slobodskikh, and S. I. Vigushina, *Vysokomol. Soedin., Ser. B* **14**, 542 (1972); *CA* **77**, 152621f.
469. M. A. Geiderikh, B. E. Davydov, N. F. Zaliznaya, and V. S. Minayeva, *Vysokomol. Soedin., Ser. A* **18**, 1264 (1976); *Polym. Sci. USSR* (*Engl. Transl.*) **18**, 1451 (1976).
470. A. A. Berlin, A. I. Kadantseva, M. A. Mukhin, and A. A. Ivanov, *Vysokomol. Soedin., Ser. A* **17**, 821 (1975); *Polym. Sci. USSR* (*Engl. Transl.*) **17**, 942 (1975).
471. M. A. Geiderikh, B. E. Davydov, N. F. Zaliznaya, and G. A. Oreshkina, *Vysokomol. Soedin., Ser. B* **11** (12), 870 (1969); *CA* **72**, 90904m.
472. N. M. Soboleva, T. F. Karpenko, and G. F. Duorko, *Dopov. Akad. Nauk Ukr. RSR, Ser. B: Geol., Khim. Biol. Nauki* **35**, 542 (1973); *CA* **79**, 105634u.
473. I. Kaneko and N. Hagihara, *J. Polym. Sci., Part B* **9**, 275 (1971).
474. D. J. Kelley, *J. Polym. Sci.* **59**, S6 (1962).
475. I. B. Rabinovich, B. V. Lebedev, L. Ya. Tsvetkova, N. S. Nametkin, S. G. Durgar'yan, V. S. Khotimskii, and Ye. G. Kiparisova, *Vysokomol. Soedin., Ser. A* **17**, 1421 (1975); *Polym. Sci. USSR* (*Engl. Transl.*) **17**, 1631 (1975).
476. N. S. Nametkin, S. G. Durgar'yan, and V. S. Khotimskii, *Vysokomol. Soedin., Ser. A* **7**, 184 (1965); *Polym. Sci. USSR* (*Engl. Transl.*) **7**, 200 (1965).
477. N. S. Nametkine, A. V. Topchiev, and S. G. Dourgarian, *J. Polym. Sci., Part C* **4**, 1053 (1963).

478. M. Stolka, J. F. Yanus, and J. M. Pearson, *Macromolecules* **9**, 376, 715, 719 (1976).
479. M. Stolka, J. F. Yanus, and J. M. Pearson, *Macromolecules* **9**, 715, 719 (1976).
480. G. F. D'Alelio and T. F. Hoffend, *J. Polym. Sci., Polym. Chem. Ed.* **5**, 1245 (1967).
481. S. Negishi and V. Tamura, *Int. Symp. Macromol. Chem., 1966* (1966).
482. J. Furukawa, T. Tsuruta, Y. Fujita, and A. Kawasiki, *J. Chem. Soc. Jpn. Ind. Chem.* **63**, 645 (1960).
483. V. G. Gasan-Zade, V. V. Mazurek, and V. P. Sklizkova, *Vysokomol. Soedin., Ser. A* **10**, 479 (1968).
484. A. Guyot, T. LeDoan, and R. Riboulet, *C. R. Hebd. Seances Acad. Sci., Ser. C* **266**, 1139 (1968); *CA* **69**, 10793z.
485. V. Jisova, M. Kolinsky, and D. Lim, *J. Polym. Sci., Polym. Chem. Ed.* **8**, 1525 (1970).
486. A. Guyot and J. Mordini, *J. Polym. Sci., Part C* **33**, 65 (1971).
487. B. Wesslen and A. Wiren, *J. Polym. Sci., Polym. Chem. Ed.* **13**, 2571 (1975).
488. R. Kray, *J. Polym. Sci.* **44**, 264 (1960).
489. J. M. Bruce and D. W. Farren, *Polymer* **4**, 407 (1963).
490. C. G. Overberger, S. Ozaki, and H. Mukamal, *J. Polym. Sci., Part B* **2**, 627 (1964).
491. R. Hart, *Bull. Soc. Chim. Belg.* **65**, 291 (1965).
492. R. C. Schulz and M. Hartman, *Makromol. Chem.* **55**, 227 (1962); R. C. Schulz and R. Stenner, *ibid.* **72**, 202 (1964).
493. C. G. Overberger and A. H. Schiller, *J. Polym. Sci., Part C* **4**, 325 (1963).
494. P. R. Thomas, G. J. Tyler, T. E. Edwards, A. T. Radcliffe, and R. C. P. Cubbon, *Polymer* **5**, 525 (1964).
495. T. Tsuruta, R. Fujio, and F. Furukawa, *Makromol. Chem.* **80**, 172 (1964).
496. T. Tsuruta and R. Tsushima, *Makromol. Chem.* **177**, 337 (1976).
497. R. Tsushima and T. Tsuruta, *Makromol. Chem.* **166**, 325 (1973).
498. R. Tsushima and T. Tsuruta, *J. Polym. Sci., Polym. Chem. Ed.* **12**, 183 (1974).
499. A. R. Lyons and E. Catterall, *Eur. Polym. J.* **7**, 839 (1971).
500. J. J. O'Malley, J. F. Yanus, and J. M. Pearson, *Macromolecules* **5**, 158 (1972).
501. G. Natta, G. Mazzanti, P. Longi, G. Dall'Asta, and F. Bernardini, *J. Polym. Sci.* **51**, 487 (1961).
502. G. Natta, G. Mazzanti, G. Dall'Asta, and P. Longi, *Makromol. Chem.* **37**, 160 (1960).
503. G. Natta, G. Mazzanti, P. Longi, G. Dall'Asta and F. Bernardini, *J. Polym. Sci.* **51**, 487 (1961).
504. C. L. Lee, J. Smid, and M. Szwarc, *Trans. Faraday Soc.* **59**, 1192 (1963).
505. A. Gourdenne, *Makromol. Chem.* **158**, 261, 271 (1972).
506. M. Tardi and P. Sigwalt, *Eur. Polym. J.* **8**, 137, 151 (1972).
507. M. Tardi, D. Rouge, and P. Sigwalt, *Eur. Polym. J.* **3**, 85 (1967).
508. M. Fortanille and P. Sigwalt, *Bull. Soc. Chim. Fr.* p. 4083 (1967).
509. C. F. Tien and T. E. Hagen-Esch, *Macromolecules* **9**, 871 (1976).
510. A. Soum, M. Fontanille, and P. Sigwalt, *J. Polym. Sci., Polym. Chem. Ed.* **15**, 659 (1977).
511. M. Fisher and M. Szwarc, *Macromolecules* **3**, 23 (1970).
512. D. Honnore, J. C. Favier, P. Sigwalt, and M. Fontanille, *Eur. Polym. J.* **10**, 425 (1974).
513. J. C. Favier, P. Sigwalt, and M. Fontanille, *Eur. Polym. J.* **10**, 717 (1974).
514. K. Tanikawa, H. Hirata, S. Kusabayashi, and H. Mikawa, *Bull. Chem. Soc. Jpn.* **42**, 2406 (1969); *CA* **71**, 102310j.
515. P. P. Spiegelman and G. Parravano, *J. Polym. Sci., Part A* **2**, 2245 (1964).
516. J. P. Schroeder, D. C. Schroeder, and S. Jotikasthira, *J. Polym. Sci., Polym. Chem. Ed.* **10**, 2189 (1972).
517. C. G. Overberger and A. M. Schiller, *J. Org. Chem.* **26**, 4230 (1961).
518. J. Boor, Jr. and A. M. T. Finch, *J. Polym. Sci., Polym. Chem. Ed.* **9**, 249 (1971).
519. E. N. Prilezhaeva and A. V. Sviridova, USSR Patent 226,150 (1968); *CA* **70**, 29,597f.

520. N. A. Plate and V. V. Mal'tsev, *Kinet. Mech. Polyreactions, Int. Symp. Macromol. Chem., Prepr. 1969* Vol. 2, p. 27 (1969); *CA* **75**, 77350v.
521. N. A. Plate, V. V. Mal'tsev, and V. P. Markina, *Vysokomol. Soedin., Ser. B* **14**, 831 (1972); *CA* **78**, 124973z.
522. N. A. Plate, V. V. Mal'tsev, S. L. Davydova, and V. A. Kargin, *Vysokomol. Soedin.* **8**, 1890 (1966).
523. Y. Miura, M. Kinoshita, and M. Imoto, *Makromol. Chem.* **146**, 69 (1971).
524. A. Konishi, *Bull. Chem. Soc. Jpn.*, **35**, 197 (1962).
525. H. Gilbert, F. F. Miller, S. J. Averill, R. F. Schmidt, F. D. Stewart, and H. L. Trumbull, *J. Am. Chem. Soc.* **76**, 1074 (1954).
526. M. Kamachi and S. Murahashi, *Polym. J.* **4**, 651 (1973).
527. S. Murahashi and M. Kamachi, *Makromol. Chem.* **119**, 232 (1968).
528. Sumitomo Chemical Co. Ltd., Japanese Patent 66/13,637 (1966); *CA* **66**, 38352s.
529. B. Yamada, R. W. Campbell, and O. Vogl, *Polym. J.* **9**, 23 (1977).
530. B. Yamada, R. W. Campbell, and O. Vogl, *J. Polym. Sci., Polym. Chem. Ed.* **15**, 1123 (1977).
531. I. Rosen, C. L. Sturm, G. H. McCain, R. M. Wilhjelm, and D. E. Hudgin, *J. Polym. Sci., Part A* **3**, 1545 (1965).
532. W. K. Busfield, *Polymer* **9**, 479 (1968).
533. S. Temple and R. L. Thornton, *J. Polym. Sci., Polym. Chem. Ed.* **10**, 709 (1972).
534. I. V. Andreeva, Y. V. Medvedev, and M. M. Koton, *Dokl. Akad. Nauk SSSR* **180**, 1075 (1969); *CA* **71**, 70963f.
535. R. C. Schulz, G. Wegner, and W. Kern, *J. Polym. Sci., Part C* **16**, 989 (1967).
536. Asahi Chem. Industry Co. Ltd., Japanese Patents 67/15,621, 67/15,622 (1967); *CA* **67**, 117,517y, 117,518t.
537. T. Kodaira, J. Stehlicek, and J. Sebenda, *Eur. Polym. J.* **6**, 1451 (1970).
538. I. Negulescu and O. Vogl, *J. Polym. Sci., Polym. Chem. Ed.* **15**, 2995 (1976).
539. O. Vogl, *J. Polym. Sci.* **46**, 261 (1960).
540. J. Furukawa, T. Saegusa, and H. Fujii, *Makromol. Chem.* **44–46**, 398 (1961).
541. O. Vogl, *J. Polym. Sci., Part A* **2**, 4607 (1964).
542. O. Vogl and W. M. D. Bryant, *J. Polym. Sci., Part A* **2**, 4633 (1964).
543. K. Hashimoto, H. Sumitomo, and S. Ohsawa, *J. Polym. Sci., Polym. Chem. Ed.* **14**, 1221 (1976).
544. I. M. Panayotov and I. B. Rashkov, *Makromol. Chem.* **154**, 129 (1972).
545. V. V. Kulakov, I. V. Kamenskii, and I. A. Andreeva, *Tr. Mosk. Khim.-Tekhnol. Inst. im. D. I. Mendeleeva* **66**, 179 (1970); *CA* **75**, 77355a.
546. Y. Kawakami, Y. Yasuda, and T. Tsuruta, *J. Macromol. Sci., Chem.* **A3**, 205 (1969).
547. Y. Yasuda, N. Kawabata, and T. Tsuruta, *J. Macromol. Sci., Chem.* **A1**, 669 (1967); **A2**, 943 (1968).
548. K. Ree and Y. Minoura, *Makromol. Chem.* **177**, 2897 (1976).
549. M. Kamachi and S. Murahasi, *Bull. Chem. Soc. Jpn.* **46**, 2890 (1973).
550. K. Hashimoto and H. Sumitomo, *J. Polym. Sci., Polym. Chem. Ed.* **7**, 1549 (1969).
551. H. Sumitomo and K. Hashimoto, *J. Polym. Sci., Polym. Chem. Ed.* **7**, 1331 (1969).
552. G. O. Cash, Jr. and J. C. Martin, U. S. Patent 3,321,441 (1967); *CA* **68**, 13684e.
553. K. Yoshida and Y. Yamashita, *Makromol. Chem.* **100**, 175 (1967).
554. Y. Yamashita and S. Nunomoto, *Makromol. Chem.* **58**, 244 (1962).
555. G. F. Pregaglia, M. Minaghi, and M. Cambini, *Makromol. Chem.* **67**, 10 (1963).
556. G. Natta, G. Mazzani, G. F. Pregaglia, and M. Binaghi, *Makromol. Chem.* **44–46**, 537 (1961).
557. G. Natta, G. Mazzani, G. F. Pregaglia, M. Binaghi, and M. Gransbini, *Makromol. Chem.* **51**, 148 (1962).
558. R. H. Hasek, R. D. Clark, E. U. Elam, and J. C. Martin, *J. Org. Chem.* **27**, 60 (1962).

559. N. F. Proshlyakova, I. F. Sanaya, and N. S. Yenikolopyan, *Vysokomol. Soedin.* **5**, 1632, 1653 (1963); *Polym. Sci. USSR* (*Engl. Transl.*) **5**, 740, 898 (1964).
560. W. I. Jrzack, L. M. Romanov, and N. S. Yenikolopyan, *Vysokomol. Soedin.* **5**, 1638 (1963); *Polym. Sci. USSR* (*Engl. Transl.*) **5**, 747 (1964).
561. L. P. Bobkova, V. S. Korsakov, L. M. Romanov, and N. S. Yenikolopyan, *Vysokomol. Soedin.* **5**, 1653, 1785 (1963); *Polym. Sci. USSR* (*Engl. Transl.*) **5**, 763, 903 (1964).
562. N. S. Enikolopyan, *J. Polym. Sci.* **58**, 1301 (1962).
563. E. Kunzel, A. Gieffer, and W. Kern, *Makromol. Chem.* **96**, 17 (1966).
564. N. Mathes and V. Jaacks, *Makromol. Chem.* **135**, 49 (1970).
565. P. I. Penchev and N. S. Enikolopyan, *Dokl. Bolg. Akad. Nauk* **26**, 235 (1973).
566. V. V. Kulakov and I. V. Kamenskii, *Tr. Mosk. Khim.-Tekhnol. Inst. im. D. I. Mendeleeva* **66**, 166 (1970); *CA* **75**, 77367f.
567. K. Hashimoto, H. Sumitomo, and S. Ohsawa, *J. Polym. Sci., Polym. Chem. Ed.* **13**, 2775 (1975).
568. S. Tagami and T. Kunitake, *Makromol. Chem.* **175**, 3367 (1974).
569. C. Aso and S. Tagami, *Macromolecules* **2**, 414 (1969).
570. K. Yamaguchi and Y. Minoura, *J. Polym. Sci., Polym. Chem. Ed.* **8**, 1571 (1970).
571. K. Kobayashi and H. Hiroshi, *J. Polym. Sci., Part B* **10**, 703 (1972).
572. D. Woehrle, *Makromol. Chem.* **160**, 83 (1972).
573. L. H. Peebles, Jr. and J. Brandrup, *Makromol. Chem.* **98**, 189 (1966).
574. I. B. Johns, *Polym. Prepr. Am. Chem. Soc., Div. Polym. Chem.* **5**, 239 (1964).
575. V. E. Shashoua, *J. Am. Chem. Soc.* **81**, 3156 (1959).
576. V. E. Shashoua, W. Sweeny, and R. F. Tietz, *J. Am. Chem. Soc.* **82**, 866 (1960).
577. G. Natta, J. D. Pietro, and M. Cambini, *Makromol. Chem.* **56**, 200 (1962).
578. T. Minami, H. Kawai, T. Agawa, and S. Komori, *Kogyo Kagaku Zasshi* **70**, 224 (1967); *CA* **68**, 40163r.
579. M. Goodman and S. C. Chen, *Macromolecules* **3**, 398 (1970).
580. M. Goodman and S. C. Chen, *Macromolecules* **4**, 625 (1971).
581. H. Yu, A. J. Bur, and L. J. Fetters, *J. Chem. Phys.* **44**, 2568 (1966).
582. A. J. Bur and L. J. Fetters, *Macromolecules* **6**, 874 (1973).
583. Y. Iwakura, K. Uno, and N. Kobayashi, *J. Polym. Sci., Polym. Chem. Ed.* **4**, 1013 (1966).
584. L. J. Fetters and H. Yu, *Polym. Prepr.* **7**, 443 (1966).
585. N. S. Schneider, S. Furusaki, and R. W. Lenz, *J. Polym. Sci., Part A* **3**, 933 (1965).
586. T. T. Patton, U. S. Patent 3,684,773 (1970); *CA* **77**, 153157w.
587. C. King, *J. Am. Chem. Soc.* **86**, 437 (1964).
588. A. Oku, T. Shono, and R. Oda, *Makromol. Chem.* **100**, 224 (1967).
589. G. C. Corfield and A. Crawshaw, *Chem. Commun.* No. 4, p. 85 (1966).
590. G. C. Corfield and A. Crawshaw, *J. Macromol. Sci., Chem.* **A5**, 3 (1971).
591. G. C. Robinson, *J. Polym. Sci., Part A* **2**, 3901 (1964).
592. J. E. Figuerulo and A. Bello, *J. Macromol. Sci., Chem.* **A3**, 311 (1969); *CA* **70**, 106921.
593. A. A. Solovyanov and K. S. Kazanskii, *Vysokomol. Soedin., Ser. A* **12**, 2114 (1970); *Polym. Sci. USSR* (*Engl. Transl.*) **12**, 2396 (1970); *CA* **73**, 131403p.
594. A. A. Solovyanov and K. S. Kazanskii, *Vysokomol. Soedin., Ser. A* **14**, 1072 (1972); *Polym. Sci. USSR* (*Engl. Transl.*) **14**, 1196 (1973); *CA* **77**, 48911q.
595. I. Cabasso and A. Zilkha, *J. Macromol. Sci., Chem.* **A8**, 1313 (1974).
596. K. S. Kazanskii, A. A. Solovyanov, and S. G. Entelis, *Eur. Polym. J.* **7**, 1421 (1971).
597. J. E. Figueruelo and D. J. Worsfold, *Eur. Polym. J.* **4**, 439 (1968).
598. C. E. H. Bawn, A. Ledwith, and N. McFarlane, *Polymer* **10**, 653 (1969); **8**, 484 (1967).
599. D. H. Richard and M. Szwarc, *Trans. Faraday Soc.* **55**, 1644 (1959).
600. C. C. Price and D. D. Carmelite, *J. Am. Chem. Soc.* **88**, 4039 (1966).

601. G. J. Stockburger and J. D. Brandner, *J. Am. Oil Chem. Soc.* **40**, 509 (1963).
602. N. N. Lebedev and Yu. I. Baranov, *Polym. Sci. USSR* (*Engl. Transl.*) **8**, 211 (1966); *Vysokomol. Soedin., Ser. A* **8**, 198 (1966).
603. B. Wojtech, *Makromol. Chem.* **66**, 180 (1963).
604. H. Kammerer and P. N. Grover, *Makromol. Chem.* **177**, 965 (1976).
605. S. Nenna and J. E. Figueruelo, *Makromol. Chem.* **176**, 3377 (1975).
606. A. Bar-Ilan and A. Zilkha, *J. Macromol. Sci., Chem. A* **4**, 1727 (1970).
607. D. Lassalle, S. Boileau, and P. Sigwalt, *Eur. Polym. J.* **13**, 587, 591, 599 (1977).
608. S. Nenna and J. E. Figueruelo, *Eur. Polym. J.* **11**, 511 (1975).
609. H. Kaemmerer and P. N. Grover, *Makromol. Chem.* **96**, 270 (1966).
610. A. A. Solovyanov and K. S. Kazanskii, *Adv. Ionic Polym., Proc. Int. Symp., 1972* p. 95 (1975).
610a. N. Oguni, S. Watanabe, M. Maki, and H. Tani, *Macromolecules* **6**, 195 (1973).
610b. N. Oguni, S. Maeda, and H. Tani, *Macromolecules* **6**, 549 (1973).
611. N. P. Doroshenko and Y. L. Spirin, *Polym. Sci. USSR* (*Engl. Transl.*) **A12**, 2812 (1970); *CA* **74**, 54245e.
612. M. A. Achon, U. S. Patent 3,598,765 (1971); *CA* **75**, 130635v.
613. J. Furukawa, T. Tsuruta, R. Sakata, T. Saegusa, and A. Kawasaki, *Makromol. Chem.* **32**, 90 (1959).
614. E. C. Steiner, R. R. Pelletier, and R. O. Trucks, *J. Am. Chem. Soc.* **86**, 4678 (1964).
615. C. C. Price and M. Osgan, *J. Am. Chem. Soc.* **78**, 4787 (1956).
616. G. A. Gladkoskii, L. P. Golovina, and V. S. Shtykh, *Vysokomol. Soedin., Ser. A* **14**, 1174 (1972); *Polym. Sci. USSR* (*Engl. Transl.*) **14**, 1315 (1972).
617. L. P. Blanchard, V. Hornof, J. Moinard, and F. Tahiani, *J. Polym. Sci., Polym. Chem. Ed.* **10**, 3089 (1972).
618. J. Lal, *J. Polym. Sci., Part B* **5**, 793 (1967).
619. N. Oguni, T. Fujita, and H. Tani, *Macromolecules* **6**, 325 (1973).
620. L. P. Blanchard, K. T. Dinh, J. Moinard, and F. Tahiani, *J. Polym. Sci., Polym. Chem. Ed.* **10**, 1353 (1972).
621. H. Sakakihara, Y. Takahashi, H. Tadokoro, N. Oguni, and H. Tani, *Macromolecules* **6**, 205 (1973).
622. N. Doddi, W. C. Forsman, and C. C. Price, *Macromolecules* **4**, 648 (1971).
623. A. Sato, T. Hirano, and T. Tsuruta, *Makromol. Chem.* **177**, 3059 (1976).
624. M. Ishimori, K. Tsukigawa, and T. Tsuruta, *Makromol. Chem.* **177**, 1221 (1976).
625. F. D. Trischler and J. Hollander, *Polym. Prepr., Am. Chem. Soc., Div. Polym. Chem.* **8**, 491 (1967).
626. M. Ito, K. Abe, T. Tsuchiya, M. Umechara, and S. Yamashiro, *Kitami Kogyo Tanki Daigaku Kenkyu Hokoku* **3**, 131 (1971); *CA* **77**, 5839e.
627. V. A. Ponomarenko, E. I. Klabunovskii, A. A. Il'chenko, and A. P. Bokovoi, *Vysokomol. Soedin., Ser. A* **11**, 1793 (1969).
628. N. Spassky, A. Pourdjavadi, and P. Sigwalt, *Eur. Polym. J.* **13**, 467 (1977).
629. A. A. Il'chenko, V. A. Ponomarenko, E. I. Klabunovskii, and L. A. Kiseleva, *Epoksidnye Monomery, Epoksidnye Smoly* p. 168 (1975); *CA* **85**, 6135z.
630. Yu. L. Spirin and N. P. Doroshenko, *Vysokomol. Soedin., Ser. A* **14**, 646 (1972); *Polym. Sci. USSR* (*Engl. Transl.*) **14**, 721 (1972).
631. I. M. Panayotov, I. V. Berlinova, and C. B. Tsvetanov, *Eur. Polym. J.* **7**, 127 (1971).
632. J. L. Villa, U. S. Patent 3,725,337 (1973); *CA* **79**, 19734v.
633. S. Boileau and P. Sigwalt, *J. Polym. Sci., Part C* **16**, 3021 (1967).
634. P. Guérin, P. Hemery, S. Boileau, and P. Sigwalt, *Eur. Polym. J.* **7**, 953 (1971); P. Hemery, S. Boileau, and P. Sigwalt, *ibid.* p. 1581.
635. G. Tersac, S. Boileau, and P. Sigwalt, *Makromol. Chem.* **149**, 153 (1971).
636. P. Guerin, S. Boileau, and P. Sigwalt, *Eur. Polym. J.* **7**, 1119 (1971).

637. T. Tsunetsugu, J. Furukawa, and T. Tueno, *J. Polym. Sci., Polym. Chem. Ed.* **9**, 3541 (1971).
638. J. C. Favier, S. Boileau, and P. Sigwalt, *Eur. Polym. J.* **4**, 3 (1968).
639. S. Boileau and P. Sigwalt, *Eur. Polym. J.* **3**, 57 (1967).
640. S. Boileau, P. Sigwalt, and N. D'Haeyer, *Bull. Soc. Chim. Fr.* p. 1054 (1968).
641. G. Tersac, S. Boileau, and P. Sigwalt, *J. Chim. Phys. Physicochim. Biol.* **65**, 1141 (1968).
642. M. Morton, R. F. Kammereck, and L. J. Fetters, *Br. Polym. J.* **3**, 120 (1971); *Macromolecules* **4**, 11 (1971).
643. P. Hemery, S. Boileau, and P. Sigwalt, *J. Polym. Sci., Part B* **13**, 49 (1975).
644. P. Hemery, S. Boileau, and P. Sigwalt, *J. Polym. Sci., Part C* **15**, 189 (1975).
645. A. D. Aliev, I. P. Solomatina, and B. A. Krentsl, *Macromolecules* **6**, 797 (1973).
646. M. Sepulchre, N. Spassky, and P. Sigwalt, *Macromolecules* **5**, 92 (1972).
647. N. Spassky, P. Dumas, M. Moreau, and J. P. Vairon, *Macromolecules* **8**, 956 (1975).
648. M. Morton and R. F. Kammereck, *J. Am. Chem. Soc.* **92**, 3217 (1970).
649. N. Mathes and V. Jaacks, *Makromol. Chem.* **142**, 209 (1971).
650. C. D. Eisenbach and R. W. Lenz, *Makromol. Chem.* **177**, 2539 (1976).
651. V. Jaacks and N. Mathes, *Makromol. Chem.* **131**, 295 (1970).
652. J. Cornibert, R. H. Marchessault, A. E. Allegrezza, Jr., and R. W. Lenz, *Macromolecules* **6**, 676 (1973).
653. H. K. Hall, Jr., *Macromolecules* **2**, 488 (1969).
654. V. Jarm and D. Fles, *J. Polym. Sci., Polym. Chem. Ed.* **15**, 1061 (1977).
655. H. K. Hall, Jr., J. B. Dence, and D. R. Wilson, *Macromolecules* **2**, 475 (1969).
656. H. K. Hall, Jr., E. P. Blanchard, Jr., and E. L. Martin, *Macromolecules* **4**, 142 (1971).
657. H. Deibig, J. Geiger, and M. Sander, *Makromol. Chem.* **145**, 123 (1971).
658. M. Ishigami, Y. Okuni, and S. Kakutan, Japanese Patent 6,900,872 (1969); *CA* **70**, 115719d.
659. B. O. Baum, U. S. Patent 3,440,227 (1969); *CA* **70**, 115720x.
660. N. V. Stamicarbon, Netherlands Patent Appl. 6,800,386 (1969); *CA* **71**, 113473v.
661. Badische Anilin- und Soda-Fabrik AG, French Patent 1,551,631 (1968); *CA* **71**, 50675k.
662. N. V. Stamicarbon, Netherlands Patent Appl. 6,802,127 (1969); *CA* **71**, 125210v.
663. V. V. Korshak, T. M. Frunze, V. I. Zaitsev, V. V. Kurashev, and T. M. Babchinitsev, *Vysokomol. Soedin., Ser. A* **12**, 416 (1970); *Polym. Sci. USSR (Engl. Transl.)* **12**, 475 (1970).
664. A. A. Kveglis and J. M. Kolyer, Ger. Offen. 2,008,117 (1970); *CA* **73**, 110504g.
665. J. Sabenda and R. Puffr, Ger. Offen 2,012,105 (1970); *CA* **74**, 32172b.
666. T. Kataoka, T. Yasumoto, and E. Naito, Japanese Patent 68/00,631 (1968); *CA* **69**, 10922v.
667. G. J. Schmitt and H. K. Reimschuessel, U. S. Patent 3,403,135 (1968); *CA* **69**, 107286z.
668. J. Sabenda and V. Kouril, *Eur. Polym. J.* **8**, 437 (1972).
669. H. E. Johnson, U. S. Patent 3,671,501 (1972); *CA* **77**, 75832w.
670. J. Roda, J. Kralicek, and K. Sanda, *Eur. Polym. J.* **12**, 729 (1976).
671. H. B. Brouns, U. S. Patent 3,345,340 (1967); *CA* **67**, 117,514v.
672. J. A. Moore and T. J. Pahls, *Angew. Makromol. Chem.* **55**, 141 (1976).
673. E. Eusebi and J. B. Nowell, U. S. Patent 3,763,077 (1973); *CA* **80**, 27877e.
674. T. Tsukamoto and H. Okamoto, Japan Kokai 73/84,898 (1973); *CA* **80**, 122250d.
675. V. N. Mikheyev, *Vysokomol. Soedin. Ser. A* **15**, 1630 (1973); *Polym. Sci. USSR (Engl. Transl.)* **15**, 1829 (1973).
676. T. Matsuo and S. Mori, U. S. Patent 3,803,101 (1974); *CA* **81**, 38119f.
677. C. E. Moyer, Jr., U. S. Patent 3,671,499 (1972); *CA* **77**, 75834y.
678. J. M. Kolyer, U. S. Patent 3,658,747 (1972); *CA* **77**, 35652f.
679. S. Zahorovsky, *Chem. Prum.* **22**, 566 (1972); *CA* **78**, 125242x.
680. T. Matsuo and S. Mori, Japan Kokai, 73/12,898 (1973); *CA* **78**, 148478t.
681. M. Matzner, L. M. Robeson, R. J. Greff, and J. E. McGrath, *Angew. Makromol. Chem.* **26**, 137 (1972).

682. P. Cefflin, J. Stehlicek, and J. Sabenda, *Eur. Polym. J.* **10**, 227 (1974).
683. G. Stea and G. B. Gechele, *Eur. Polym. J.* **6**, 233 (1970).
684. R. Z. Greenley, J. C. Stauffer, and J. E. Kurz, *Macromolecules* **2**, 561 (1969).
685. H. Sekiguchi and B. Coutin, *Macromol. Synth.* **6**, 57 (1977); *CA* **86**, 171884w.
686. H. Sekiguchi, P. Rapacoulia, and B. Coutin, *Nuova Chim.* **49**, 32 (1973); *CA* **78**, 98104v.
687. H. Sekiguchi and B. Coutin, *J. Polym. Sci., Polym. Chem. Ed.* **11**, 1601 (1973).
688. N. Yoda and A. Miyake, *J. Polym. Sci.* **43**, 117 (1960).
689. H. Tani and Y. Konomi, *J. Polym. Sci., Polym. Chem. Ed.* **4**, 301 (1966); **6**, 2281, 2295 (1968); **7**, 2255 (1969).
690. C. D. Eisenbach and R. W. Lenz, *Macromolecules* **9**, 227 (1976).
691. J. Sabenda, J. Hauer, and J. Biros, *J. Polym. Sci., Polym. Chem. Ed.* **14**, 2357 (1976).
692. Farbwerke Hoechst, AG, French Patent 1,517,147 (1968); *CA* **70**, 88440x.
693. E. Schmidt, *Angew. Makromol. Chem.* **14**, 185 (1970).
694. H. Sekiguchi, P. Rapacoulia, and B. Coutin, French Patent 2,138,228 (1973); *CA* **79**, 92858t.
695. D. E. Sargent, U. S. Patent 3,647,764 (1972); *CA* **76**, 154423w.
696. M. Kushova, J. Kralicek, and J. Roda, *Eur. Polym. J.* **13**, 665 (1977).
697. H. Sekiguchi, P. Rapacoulia, and B. Coutin, *J. Polym. Sci., Part C* **42**, 51 (1973).
698. G. Schirawski, *Makromol. Chem.* **161**, 57 (1972).
699. J. Budin, *Collect. Czech. Chem. Commun.* **41**, 220 (1976).
700. S. Barzakay, M. Levy, and D. Vofsi, *J. Polym. Sci., Polym. Chem. Ed.* **4**, 2211 (1966).
701. S. Barzakay, M. Levy, and D. Vofsi, *J. Polym. Sci., Part B* **3**, 601 (1965).
702. N. A. Mukhitdinova, *Sin. Modif. Sin. Polim.* p. 105 (1971); *CA* **77**, 52639t.
703. F. P. Sidelkovskaya, N. A. Mukhitdinova, M. G. Zelenskaya, and V. A. Ponomarenko, *Fiziol. Opt. Akt. Polim. Veshchestva, Tr. Vses. Simp., 2nd, 1969* p. 50, (1971); *CA* **77**, 165112r.
704. N. A. Mukhitdinova, T. V. Markman, and M. A. Askarov, *Sint. Vysokomol. Soedin.* p. 17, (1972); *CA* **79**, 5649j.
705. H. Sumitomo, K. Hashimoto, and M. Ando, *J. Polym. Sci., Part B* **11**, 635 (1973).
706. W. R. Hertler, W. H. Sharkey, and B. C. Anderson, *Macromolecules* **9**, 523 (1976).
707. H. Zweifel and T. Voelker, *Angew. Chem., Int. Ed. Engl.* **11**, 342 (1972).
708. H. Zweifel, J. Loeliger, and T. Voelker, *Makromol. Chem.* **153**, 125 (1972).
709. S. N. Bhadani and J. Prasad, *Makromol. Chem.* **178**, 1651 (1977).
710. D. Braun and J. Pomakis, *Makromol. Chem.* **175**, 1411 (1974).
711. I. Schopov, *Makromol. Chem.* **137**, 293 (1970).
712. M. L. Hallensleben and H. Wurm, *Eur. Polym. J.* **9**, 919 (1973).
713. M. Oya, K. Uno, and Y. Iwakura, *Bull. Chem. Soc. Jpn.* **43**, 1788 (1970); *CA* **73**, 56475k.
714. B. M. Abo-El Khair, T. Komoto, and T. Kawai, *Makromol. Chem.* **177**, 2481 (1976).
715. T. Komoto, B. M. Abo-El Khair, K. Maeda, M. Oya, and T. Kawai, *Makromol. Chem.* **177**, 2481, 2505 (1976).
716. H. R. Kricheldorf and K. Bösinger, *Makromol. Chem.* **177**, 1243 (1976).
717. H. R. Kricheldorf, *Makromol. Chem.* **175**, 3325 (1974).
718. E. Hasegawa, T. Shibata, H. Takeuchi, and E. Tsuchida, *Eur. Polym. J.* **13**, 669 (1977).
719. H. Sekiguchi and G. Froyer, *J. Polym. Sci., Part C* **52**, 157 (1975).
720. T. Komoto, K. Y. Kim, M. Oya, and T. Kawai, *Makromol. Chem.* **175**, 283, 301 (1976); **168**, 261 (1973).
721. Y. Iwakura, K. Uno, and M. Oya, *J. Polym. Sci., Polym. Chem. Ed.* **5**, 2867 (1967).
722. M. Goodman and J. Hutchison, *J. Am. Chem. Soc.* **88**, 3627 (1966).
723. H. G. Elias and H. B. Buehrer, *Makromol. Chem.* **140**, 21 (1970).
724. H. R. Kricheldorf, *Makromol. Chem.* **175**, 3325 (1974).
725. D. T. Hurd, R. C. Osthoff, and M. L. Corrin, *J. Am. Chem. Soc.* **76**, 249 (1954).
726. W. T. Grubb and R. C. Osthoff, *J. Am. Chem. Soc.* **77**, 1405 (1955).

727. V. P. Davydova, Z. S. Lebedeva, and I. A. Gryaznykh, *Vysokomol. Soedin., Ser. A* **9**, 2226 (1967); *Polym. Sci. USSR (Engl. Transl.)* **9**, 2518 (1967).
728. M. Morton and E. E. Bostick, *J. Polym. Sci., Part A* **2**, 523 (1964).
729. M. Morton, M. A. Deisz, and E. E. Bostick, *J. Polym. Sci., Part A* **2**, 513 (1964).
730. J. B. Carmichael and R. Winger, *J. Polym. Sci., Part A* **3**, 971 (1965).
731. G. D. Cooper and J. R. Elliott, *J. Polym. Sci., Polym. Chem. Ed.* **4**, 603 (1966).
732. R. L. Ostrozynski, *Polym. Prepr. Am. Chem. Soc., Div. Polym. Chem.* **8**, 474 (1967).
733. V. P. Davydova and Z. S. Lebedeva, *Vysokomol. Soedin., Ser. B* **10**, 401 (1968); *CA* **69**, 52815p.
734. D. R. Weyenberg, D. E. Findlay, J. Cekada, Jr., and A. E. Bey, *Polym. Prepr. Am. Chem. Soc., Div. Polym. Chem.* **7**, 562 (1966); *J. Polym. Sci., Part C* **27**, 27 (1969).
735. Z. Laita and M. Jelinek, *Vysokomol. Soedin.* **4**, 1739 (1962); *Polym. Sci. USSR (Engl. Transl.)* **4**, 535 (1963).
736. C. L. Lee, C. L. Frye, and O. K. Johannson, *Polym. Prepr., Am. Chem. Soc., Div. Polym. Chem.* **10**, 1361 (1969).
737. C. L. Lee and O. K. Johannson, *J. Polym. Sci., Polym. Chem. Ed.* **14**, 729, 743 (1976).
738. J. G. Zilliox, J. E. L. Roovers, and S. Bywater, *Macromolecules* **8**, 573 (1975).
739. H. J. Hoelle and B. R. Lehnen, *Eur. Polym. J.* **11**, 663 (1975).
740. J. Chojnowski and M. Mazurek, *Adv. Ionic Polym., Proc. Int. Symp., 1972* p. 71 (1975); *CA* **85**, 21942d.
741. E. E. Bostick, *Polym. Prepr., Am. Chem. Soc., Div. Polym. Chem.* **10**, 877 (1969).
742. Yu. A. Larionova, A. I. Ponomarev, Yu. A. Yuzhelevskii, and A. I. Klebanskii, *Vysokomol. Soedin., Ser. B* **16**, 298 (1974); *CA* **81**, 106069s.
743. Yu. A. Yuzhelevskii, V. V. Pchelintsev, and Ye. G. Kagan, *Vysokomol. Soedin., Ser. A* **15**, 1602 (1973); *Polym. Sci. USSR (Engl. Transl.)* **15**, 1795 (1973).
744. Yu. A. Yuzhelevskii, E. B. Dmokhovskaya, A. L. Klebanskii, and N. V. Kozlova, *Vysokomol. Soedin., Ser. A* **11**, 432 (1969); *Polym. Sci. USSR (Engl. Transl.)* **11**, 489 (1969).
745. K. A. Andrianov, I. I. Tverdokhlebova, S. S. Pavlova, T. A. Layina, and I. M. Petrova, *Vysokomol. Soedin., Ser. A* **18**, 1117 (1976); *Polym. Sci. USSR (Engl. Transl.)* **18**, 1284 (1977).
746. Yu. A. Yuzhelevskii, Ye. G. Kagan, N. P. Timofeyeva, T. D. Doletskaya, and A. L. Klebanskii, *Vysokomol. Soedin., Ser. A* **13**, 183 (1971); *Polym. Sci, USSR (Engl. Transl.)* **13**, 208 (1971).
747. S. N. Borisov, T. V. Kurlova, Yu. A. Yuzhelevskii, E. A. Chernyshev, and N. N. Fedoseeva, *Prom. Sin. Kauch. Nauch.-Tekh. Sb.* **1**, 7 (1970); *CA* **75**, 88985g.
748. N. Saghian and D. Gertner, *J. Macromol. Chem.* **A9**, 597 (1975).
749. Yu. A. Yuzhelevskii, Ye. G. Kagan, E. V. Kogan, A. L. Klebanskii, and N. N. Nikiforova, *Vysokomol. Soedin., Ser. A* **11**, 1539 (1969); *Polym. Sci. USSR (Engl. Transl.)* **11**, 1745 (1969).
750. K. A. Andrianov, V. Ya. Kavalenko, and V. A. Dorofeev, *Tr. Inst. – Mosk. Inst. Tonkoi Khim. Tekhnol.* **2**, 47 (1972); *CA* **81**, 121162k.
751. K. A. Andrianov and A. B. Zachernyuk, *Vysokomol. Soedin. Ser. B* **16**, 307 (1974); *CA* **81**, 92010b.
752. N. S. Nametkin, V. M. Vdovin, V. A. Poletaev, N. N. Alekhin, and M. B. Sergeeva, Deposited Publ. 1973 VINITI 5798-73 (1976); *CA* **85**, 63388h.
753. W. G. Davies and H. V. A. Beedle, Ger. Offen. 2,542,425 (1976); *CA* **85**, 22248u.
754. B. Suryanarayanan, B. W. Peace, and K. G. Mayhan, *J. Polym. Sci., Polym. Chem. Ed.* **12**, 1089, 1109 (1974).
755. W. A. Piccoli, G. G. Haberland, and R. L. Merker, *J. Am. Chem. Soc.* **82**, 1883 (1960).
756. J. Chojnowski and M. Mazurek, *Makromol. Chem.* **176**, 2999 (1975).
757. H. K. Hall, Jr., *Macromolecules* **4**, 139 (1971).

758. O. Vogl and A. C. Knight, *Macromolecules* **1**, 315 (1968).
759. K. Soga, Y. Tazuke, S. Hosoda, and S. Ikeda, *J. Polym. Sci., Polym. Chem. Ed.* **15**, 219 (1977).
760. G. Kapienis and S. Penczek, *J. Polym. Sci., Polym. Chem. Ed.* **15**, 371 (1977).

Chapter

3 The Initiation Reaction

Polymerization reactions involving the carbon–carbon double bond can be initiated principally by three types of initiators: (i) alkali metals, (ii) aromatic complexes of alkali metals, and (iii) organoalkali compounds, mainly organolithium. (Other types of monomers, e.g., heterocyclics and heterounsaturates, can also be polymerized by weaker bases, e.g., hydroxides and alkoxides.) These three types of initiator operate by different mechanisms which are discussed below.

1. Aromatic Complexes of Alkali Metals

a. Vinyl Monomers

As mentioned previously, the real nature of the anionic mechanism of polymerization became apparent only in the case of the homogeneous systems, specifically those involving the use of aromatic complexes of alkali metals. The latter species, such as the well-known sodium naphthalene, are radical-anions which have been known for some time but whose nature has only been elucidated in more recent years. Thus Schlenk *et al.*,[1] as early as 1914, showed that the alkali metals can react with higher aromatics, in the presence of ether solvents, to form colored solutions *with no evolution of hydrogen*. This was confirmed in later work by Schlenk and Bergmann.[2] The actual structure of these complexes, however, was established only by means of the more modern techniques of electron spin resonance by Lipkin, Weissman, and co-workers.[3,4] They showed that these species are *radical-anions* which have an extra electron in the lowest unoccupied π orbital and that the solvent plays an especially important role in assisting the transfer of the electron from the sodium to the naphthalene and in stabilizing the complex through interorbital exchanges with the electrons available from the oxygen of the ether solvent. Hence these complexes are entirely different from compounds such as naphthylsodium ($NaC_{10}H_7$), where the sodium has actually displaced one of the hydrogen atoms.

The formation of complexes such as sodium naphthalene can be depicted by the following equation:

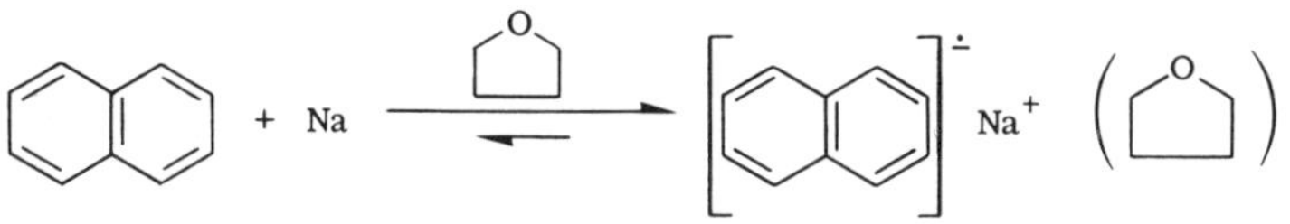
(3.1)

Here the H_4-furan represents a typical solvating ether which is effective in the formation and stabilization of the sodium naphthalene complex. Equation (3.1) also indicates that this is an equilibrium process, greatly favoring the complex formation. Weissman and associates also showed that the stability of these complexes depended on the electron affinity of the hydrocarbons, and they actually measured the equilibrium constant of the electron transfer between various higher aromatics, e.g.,

$$[\text{naphthalene}]^{\dot{-}} + \text{phenanthrene} \overset{K_e}{\rightleftharpoons} \text{naphthalene} + [\text{phenanthrene}]^{\dot{-}} \quad (3.2)$$

This enabled them to set up an "electron affinity" scale between these hydrocarbons, in the following order:

anthracene > phenanthrene > naphthalene > biphenyl.

The first use of such aromatic complexes as initiators of polymerization was apparently due to Scott[5] in 1939, who found that the greenish-blue solution of sodium naphthalene (or biphenyl) in dimethoxyethane was effective in initiating a very rapid polymerization of styrene. He also suggested that these colored solutions could polymerize other conjugated monomers, such as butadiene and isoprene. Scott and co-workers[5a] actually inferred that these complexes are dicarbanionic, since they obtained dicarboxylic salts from the usual reaction with carbon dioxide.

Sodium naphthalene was later studied as a polymerization initiator by Whitby and Stephens[6] who were able to polymerize butadiene and isoprene to relatively low molecular weight. They noted that brightly colored solutions were obtained, red for styrene, orange for dienes, but that the colors faded slowly, presumably because of atmospheric impurities.

Szwarc, Levy, and Milkovich,[7] by working under much more rigorous conditions, were able to demonstrate the "living" nature of the polymerization of styrene by sodium naphthalene, i.e., the complete absence of any noticeable termination or transfer reactions. They also pointed out that the initiation step in this case was a rapid "electron transfer" step from the naphthalene to the styrene, as follows:

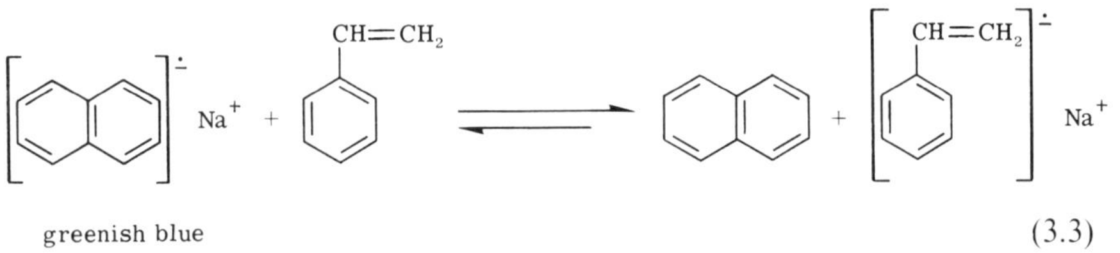

(3.3)

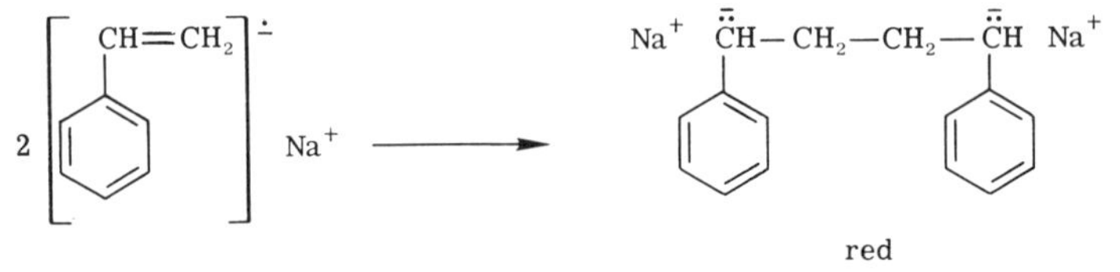

(3.4)

The dianion formed in reaction (3.4) from the rapid coupling of the styrene radical-anions is of course capable of propagating a chain of styrene units. Both reactions (3.3) and (3.4) occur quite rapidly, so that the initiation step is completed long before chain growth occurs to any extent. Under these conditions, therefore, the following situation results:

(i) The molecular weights can be predicted from simple stoichiometric consideration, i.e.,

$$M_s = \text{Wt of monomer in g/0.5 mol of initiator} \qquad (3.5)$$

(ii) The molecular weight distribution would, on theoretical grounds, be of the Poisson type (see Eq. (1.8)).

In addition, these systems also demonstrated the synthetic possibilities inherent in such living polymers, i.e., the formation of true block copolymers by sequential addition of monomers, and the synthesis of polymers with terminal functional groups. Both of these have already been referred to in Chapter 1.

All of the above conclusions became possible only because of the *homogeneous* nature of these systems, since, as will be seen later, the same type of electron-transfer initiation presumably occurs with the alkali metals as well, but is obscured by their heterogeneous character.

Any detailed study of the kinetics of initiation reactions with these aromatic alkali complexes is rendered difficult by their high speed and complexity. Hence little or no information of this type can be found in the literature. It is interesting, in this connection, to consider the nature of the actual electron transfer step (Eq. (3.3)) between the naphthalene radical-anion and the styrene. This is shown as an equilibrium reaction, and would be expected to show a strong preference to go *from right to left* in view of the markedly greater stability expected for the naphthalene radical-anion from resonance consideration. Hence the transfer of an electron from the naphthalene radical-anion to styrene represents a highly unfavorable process, until it is taken into consideration that the relatively unstable styrene radical-anion can *readily couple*. This decreases the concentration of this species to the point where reaction (3.3) proceeds forward virtually to completion.

It is interesting to note in this connection, therefore, that the electron transfer process is not governed solely by "electron affinity" considerations, as illustrated by the equilibrium shown in Eq. (3.2), but is also a function of the equilibrium concentration of the particular radical-anion, which may be strongly affected by the possibility of "side reactions," e.g., coupling of radical-anions to stable dianions. The latter is apparently favored in the case of styrene.

It also bears mentioning, at this point, that reactions of the type shown by Eqs. (3.1) and (3.3) can also lead to transfer of *two* electrons, and there is evidence that this actually occurs in the case of initiation by the alkali metals, as will be seen later. However, this phenomenon would also be governed by the effect of concentration, and would therefore not be expected to occur to any extent in the case of the soluble aromatic complexes, where concentrations of radical-anions are of the order of 10^{-3} *M*, leading to rapid coupling (radical termination) processes.

b. HETEROCYCLICS

Although only a few cases have been reported of the polymerization of heterocyclics by the aromatics complexes of alkali metals, these are of special interest since they illustrate a different mechanism of initiation. Presumably because of the polarization of the heteroatomic bond in the monomer, the initial step is a direct nucleophilic attack by the anion rather than an electron transfer step. This was shown first[8] in the case of the cyclic siloxane tetramer octamethylcyclotetrasiloxane, where the following sequence of reactions apparently occurs:

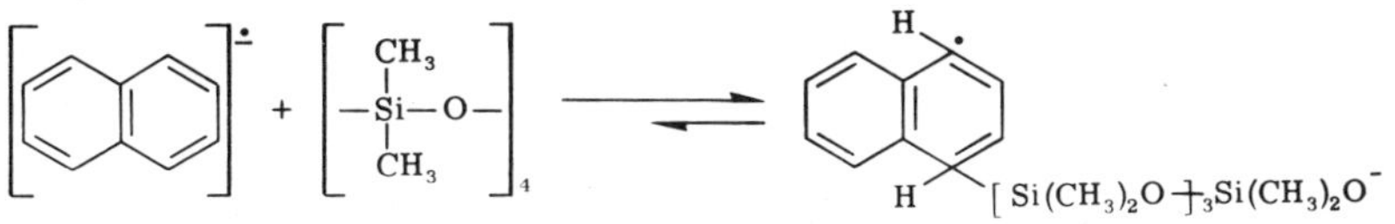

I (3.6)

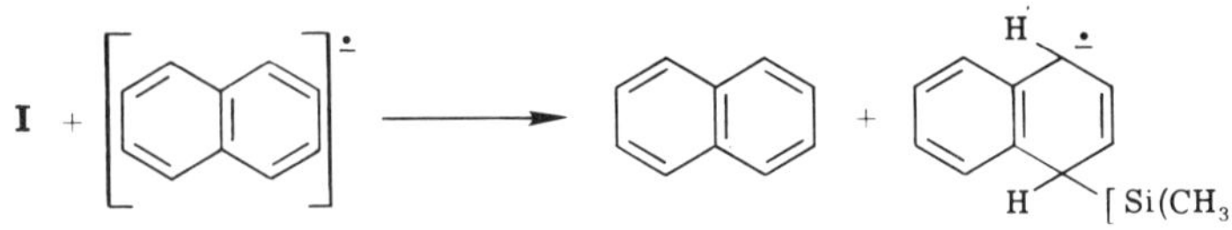

II (3.7)

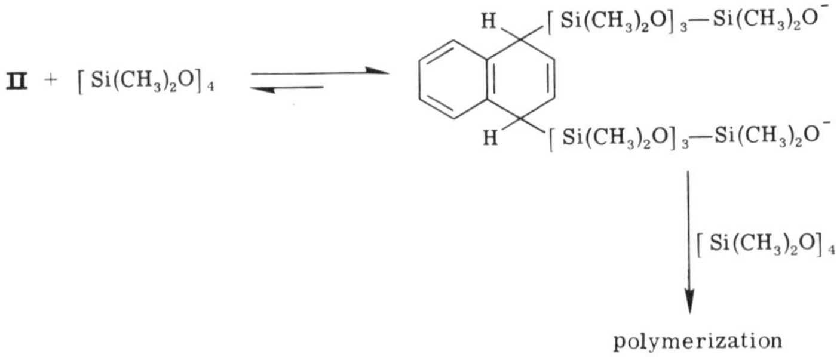

(3.8)

Although only the para isomer is shown as the final product in reaction (3.8), it is understood that all possible isomers involving the dihydronaphthalene nucleus can occur. The above mechanism was supported by evidence obtained from ultraviolet spectroscopy, which clearly showed the presence of a dihydronaphthalene residue attached to the polysiloxane. The proposed electron-transfer reaction (3.7) was also supported by the presence of a persistent faint pink color in the solution, which could be discharged by addition of water, and which was presumably due to the carbanion shown as structure **II**. The persistence of this color would be expected because of the known equilibrium character of the ring-opening reaction (3.8).

An alternate mechanism would involve the coupling of the free radicals (**I**), but this would not be favored on steric grounds and no evidence was found for the presence of a dinaphthalene structure. A similar anionic mechanism was also later proposed[9] for the polymerization of ethylene oxide by sodium naphthalene. In this, a *fleeting* pink color was noticed, presumably because of the "nonequilibrium" nature of the ring-opening reaction, i.e., a very low equilibrium concentration of the cyclic monomer.

Dudek[10] and Morton carried out an extensive study of the kinetics of polymerization of ethylene oxide by these aromatic complexes of the alkali metals. The initiation step here too was found to involve direct anionic attack. The propagation kinetics will be discussed in Chapter 4.

Another heterocyclic whose polymerization by the naphthalene complexes was studied in some detail is propylene sulfide. In the initial work[11] it was found that sodium naphthalene apparently leads to the formation of a dianion with this monomer and that the molecular weight of the polymer can be predicted by Eq. (3.5), as in the case of styrene. The distribution of molecular weights was also found to be quite narrow ($M_w/M_n \sim 1.1$–1.2), as expected for a termination-free polymerization.

Later work by Sigwalt and associates[11a] showed conclusively that, unlike the case of the cyclic oxides described above, the polymerization of propylene sulfide involved an initiation by electron transfer rather than by direct anionic attack. They proposed two alternate mechanisms for the initiation reaction, thus:

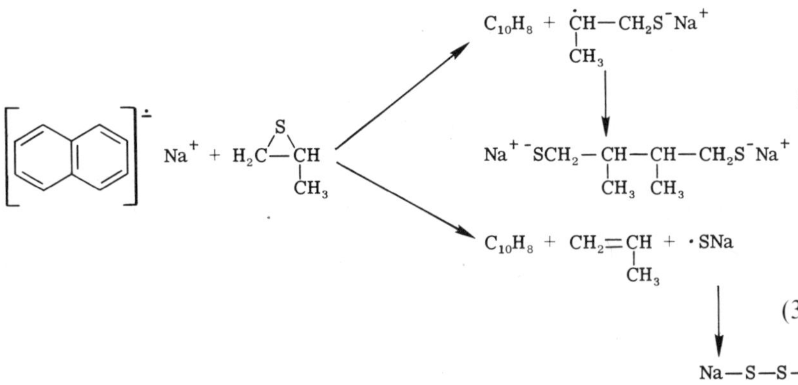

(3.9)

(3.10)

These mechanisms were supported by the fact that no naphthalene residues were found in the polymer and that the free naphthalene could be recovered. In addition, the presence of propene was also detected, but since no quantitative data were obtained, some participation of reaction (3.9) could not be ruled out.

2. The Alkali Metals

The clear-cut evidence for initiation by an electron transfer mechanism, which was provided by the aromatic complexes of the alkali metals, also suggested a similar mechanism for the well-known polymerization of dienes and styrene by alkali metals such as sodium or potassium. Thus the initiation step in the sodium-catalyzed polymerization of butadiene could be written as follows:

$$\mathrm{Na} + \mathrm{CH_2{=}CH{-}CH{=}CH_2} \rightarrow [\mathrm{CH_2{=}CH{-}CH{=}CH_2}]^{\dot{-}}\,\mathrm{Na^+} \qquad (3.11)$$

$$2[\mathrm{CH_2{=}CH{-}CH{=}CH_2}]^{\dot{-}}\,\mathrm{Na^+} \rightarrow \mathrm{Na^+}\ {}^{-}\mathrm{CH_2{-}CH{=}CH{-}CH_2{-}CH_2{-}CH{=}CH{-}CH_2^-}\ \mathrm{Na^+} \qquad (3.12)$$

and/or

$$[\mathrm{CH_2{-}CH{=}CH{-}CH_2}]^{\dot{-}}\,\mathrm{Na^+} + \mathrm{Na} \rightarrow \mathrm{Na^+}\ {}^{-}\mathrm{CH_2{-}CH{=}CH{-}CH_2^-}\ \mathrm{Na^+} \qquad (3.13)$$

Thus Eq. (3.11) represents an electron transfer from the sodium atom to the π

orbital of the conjugated diene (only one of the 1,4 resonance forms is shown but, in fact, these polymerizations initiated by sodium lead to a predominance of 1,2-addition). Equations (3.12) and (3.13) represent the two alternative reactions of the initial radical-anions, i.e., radical coupling or a second electron transfer. The choice between these two depends on the reaction conditions. Thus, as was seen in the case of soluble aromatic complexes such as sodium naphthalene, the high concentration of radical anions formed from the monomer would favor radical coupling. On the other hand, in the case of the relatively slow electron transfer reaction which occurs at the sodium metal surface, one must consider the possibility of a second electron transfer step between the low concentration of radical-anions and the sodium interface where they are generated.

As a matter of fact, the occurrence of two electron transfer steps is supported by the work of Ziegler[12] and of Robertson and Marion[13] on the sodium polymerization of butadiene. In both studies evidence of the formation of a disodium adduct of butadiene was apparently found. Hence it appears that a strongly electropositive metal such as sodium is capable of transferring a second electron to the initial radical-anion. In their work on initiation by the less electropositive lithium metal O'Driscoll and Tobolsky have even suggested that the initial radical-anion has a sufficiently long lifetime to initiate *both* anionic and free radical polymerization before becoming converted to a dianion. This conclusion was based on their findings that lithium metal was able to copolymerize a mixture of styrene and methyl methacrylate, while a lithium alkyl only formed poly(methyl methacrylate) from the same mixture. Hence they proposed that the copolymer was formed by radical initiation by the radical-anion while the poly(methyl methacrylate) resulted from simultaneous initiation by the anion part of the radical anion. The resulting polymer should then be a *block* copolymer, consisting of a block of poly(methyl methacrylate) attached to a block of the random copolymer.

Although this was an attractive hypothesis, it was later called into question by Overberger and Yamamoto,[14] who analyzed the polymer formed in the above system by NMR and showed it to consist of a *block* of polystyrene attached to a *block* of poly(methyl methacrylate). To explain this phenomenon, they invoked a special homopolymerization of styrene monomer which is preferentially adsorbed on the surface of the lithium, and which, after reaching a certain chain length, becomes detached from the lithium and then initiates the homopolymerization of the methyl methacrylate. (The exclusive homopolymerization of methyl methacrylate by organolithium in the presence of styrene had previously been demonstrated.[15])

This hypothesis was also supported by the work of Korotkov *et al.*,[16] who showed that methyllithium also led to similar results in the polymerization of mixtures of styrene and methyl methacrylate. Since initiation by methyllithium does *not* involve electron transfer, there should be no possibility of free radical formation in that case, and hence no chance of radical-initiated polymerization. However, since methyllithium is not readily soluble, it does involve a heterogeneous initiation at its surface, similar to the situation prevailing in the case of lithium metal. Hence the weight of evidence seems to favor heterogeneous initiation rather than free radicals as the primary cause of this phenomenon.

3. Organolithium Initiation

a. Vinyl Monomers

Of the organoalkali compounds which can initiate polymerization of vinyl (and other) monomers the most versatile are the organolithium type, since these are soluble in both polar and nonpolar solvents. (The higher organoalkali compounds generally require ethers as solvents.) Furthermore, they can be used under conditions which avoid undesirable side reactions, which are so prevalent with the more basic metals.

These initiators differ from the previous two groups in operating by a direct anionic (nucleophilic) attack rather than by an electron transfer mechanism. This also leads to a mono-functional chain-growth reaction, which results in a better control of molecular weight distribution than the dianionic propagation involved in initiation by electron transfer. Here the stoichiometry is also very simple since each initiator molecule presumably generates a polymer chain, i.e.,

$$M_s = \text{Wt of monomer (grams)/moles of initiator} \qquad (3.14)$$

In those cases, e.g., styrene and the dienes, where there is an absence of any noticeable termination or transfer reactions, it should be possible to measure directly and separately the kinetics of the initiation and propagation steps of the living polymerization. It is therefore not surprising that a considerable number of studies have been devoted to these questions. In principle, these systems should represent the simplest possible polymerization reactions for kinetic studies because of the "permanence" of the growing chain. In other words, the initiation step should be a simple bimolecular reaction between an initiator molecule and the monomer, while the propagation step should likewise involve a simple bimolecular reaction between the monomer and the growing chain. Unfortunately, the situation has rarely been that simple. Thus the kinetics of both the initiation and the propagation steps have invariably been found to be *first order with respect to the monomer concentration*, irrespective of solvent type, temperature, or other variables. However, both of these steps have been found to show various *fractional order dependencies on the initiator concentration*, depending on the solvent used and even on the monomer concentration. This indicates a much more complex mechanism than at first supposed.

Most of the kinetic studies of organolithium initiation have been concerned with styrene (and some of its derivatives) and the 1,3-dienes, since these can be made to be relatively free of side reactions (and also because these monomers are of greatest practical interest). However, even in these cases, the initiation

TABLE 3.1 Kinetics of Initiation by Butyllithium Isomers in Hydrocarbon Solvents[a]

Initiator	Monomer	Solvent	T(°C)	Method[b]	Reaction order	References
n-C_4H_9Li	Styrene	Benzene	25	UV	0.33	17
		Benzene	30	UV	0.167	18
		Cyclohexane	40	UV	0.5–1.0	19
		Cyclohexane	50	GC	1.0	20
		Toluene	50	GC	1.0	20
	2,4-Dimethyl-styrene	Benzene	25	UV	0.19	21
	2,6-Dimethyl-styrene	Benzene	25	UV	0.24	21
	m-Fluorostyrene	Hexane	25	UV	0.57	21
	Butadiene	n-Hexane	50	GC	1.0	20
		Toluene	50	GC	1.0	20
		Cyclohexane	40	UV	0.5–1.0	19
		Cyclohexane	50	GC	1.0	20
	Isoprene	Cyclohexane	30	UV	0.5–1.0	22
		Cyclohexane	50	GC	1.0	20
		n-Hexane	50	GC	1.0	20
		Toluene	50	GC	1.0	20
s-C_4H_9Li	Styrene	Benzene	30	UV	0.25	23
		Cyclohexane	40	UV	1.4	23
		Cyclohexane	50	GC	1.0	20
	Butadiene	Benzene	30	IR	0.9	24
		Cyclohexane	50	GC	1.0	20
	Isoprene	Cyclohexane	50	GC	1.0	20
		Benzene	30	UV	0.25	23
		Cyclohexane	30	UV	0.75	23
		Cyclohexane	25	GC	0.66	25
		n-Hexane	30	UV	0.70	26
t-C_4H_9Li	Styrene	Cyclohexane	50	GC	1.0	20
	Butadiene	Cyclohexane	50	GC	1.0	20
	Isoprene	Cyclohexane	50	GC	1.0	20
		Cyclohexane	25	GC	0.2–0.7	25
i-C_4H_9Li	Styrene	Cyclohexane	50	GC	1.0	20
	Isoprene	Cyclohexane	50	GC	1.0	20

[a] Initiator concentrations were between 10^{-4} and 5×10^{-3} M, except for Ref. 17, in which the concentration was 1.29 M.

[b] Here UV is ultraviolet spectroscopy, IR infrared spectroscopy, and GC gas chromatography.

reaction could generally only be studied in nonpolar, i.e., hydrocarbon, media, because the polar solvents such as ethers can react directly with many of the initiators. Table 3.1 contains a fairly comprehensive compilation of available kinetic data on the initiation reaction between various butyllithiums and styrene, butadiene, or isoprene in several different hydrocarbon solvents. The one striking feature of this table is the complete lack of agreement between the results of different workers, both in *different* systems, and even using the *same*

TABLE 3.2 Association of Organolithium in Hydrocarbon Solvents

Compound	Solvent	Concentration range, *M*	Association number, *n*	Method[a]	References
C_2H_5Li	Benzene	–	~6	FP	27
	Benzene	0.03–0.4	~6	FP	28
	Benzene	0.03–0.4	4.5–6.0	FP	29
	Benzene	0.02–0.23	6.07 ± 0.35	FP	30
	Cyclohexane	0.02–0.10	5.95 ± 0.3	FP	31
	Benzene	0.006–0.19	6.1 ± 0.18	FP	32, 33
	Cyclohexane	0.006–0.08	6.0 ± 0.12	FP	32, 33
i-C_3H_7Li	Cyclohexane	0.004–0.02	4.0 ± 0.08	FP	32, 33
	Cyclohexane	>0.02	>4.0	FP	32, 33
n-C_4H_9Li	Benzene	0.5–3.4	6.25 ± 0.06	I	34
	Benzene	–	~7	BP	35
	Benzene	0.002–0.6	6.0 ± 0.12	FP	32, 33
	Cyclohexane	0.4–3.3	6.17 ± 0.12	I	34
s-C_4H_9Li	Benzene	0.17–0.5	4.13 ± 0.05	FP	23
	Cyclohexane	0.11–0.4	4.12 ± 0.09	FP	23
t-C_4H_9Li	Benzene	0.05–0.18	3.8 ± 0.2	BP	36
	Benzene	0.26–0.66	4.0 ± 0.04	FP	32, 33
	Cyclohexane	0.0005–0.3	4.0 ± 0.05	FP	32, 33
	n-Hexane	0.05–0.23	4.0 ± 0.2	BP	36
n-$C_5H_{11}Li$	Benzene	0.23–2.32	6.00 ± 0.09	VP	37
n-$C_8H_{17}Li$	Benzene	0.19–2.32	5.953 ± 0.016	VP	37
$(CH_3)_3SiCH_2Li$	Benzene	0.03–1.1	4.0 ± 0.03	FP	32, 33
	Benzene	0.6–2.78	4.0 ± 0.2	BP	38
	Benzene	0.06–0.49	4.0 ± 0.11	FP	39
	2-Methylpentane	0.2–1.2	3.9 ± 0.2	BP	38
	Cyclohexane	0.002–0.05	6.0 ± 0.18	FP	32, 33
$C_6H_5CH_2Li$	Benzene	0.0072–0.036	2.2 ± 0.3	FP	32, 33
3-Butenyllithium	Cyclopentane	0.77–5.06	6.1 ± 0.8	VP	40
3-Neopentyl-allyllithium	Benzene	0.0495–0.33	2.14–3.17[b]	FP	41
Menthyllithium	Benzene	0.1 and 0.38	2.17, 1.95	FP	42
	Cyclohexane	0.23 and 0.29	1.93, 2.04	FP	42

[a] Here FP is freezing point depression, VP vapor pressure depression, BP boiling point elevation, and I isopiestic distillation.

[b] Association number decreased as concentration decreased.

systems. There seems to be a general trend indicating a reaction order of *less than unity*, but no agreement on how much less.

Two reasons can be logically advanced to explain the glaring discrepancies demonstrated in Table 3.1. Either the data are unreliable because the reaction conditions did not assure the rigorous exclusion of impurities (a discussion of the experimental techniques required for such measurements has recently been published[26a]) or the reaction orders may actually vary with the different concentration ranges of initiator and monomer used by the various authors.

The prevalence of fractional orders of dependence of initiation rate on initiator concentration has tempted some authors to relate this phenomenon to the well-known tendency of organolithium compounds to form associated complexes. Many such compounds have been analyzed for association in solution, and Table 3.2 contains a listing of most of the results. It can be seen that alkyllithiums can often associate in groups of four or six, depending on their structure, and this has led some workers in this field to relate the kinetics of initiation to the state of association of the initiator by the following type of relation:

$$(\mathrm{RLi})_n \overset{K_e}{\rightleftharpoons} n\mathrm{RLi} \tag{3.15}$$

$$\mathrm{RLi} + \mathrm{M} \rightarrow \mathrm{RMLi} \overset{\mathrm{M}}{\rightarrow} \cdots \tag{3.16}$$

where RLi represents the organolithium initiator and M is the monomer. Equations (3.15) and (3.16) are based on the concept that only the *unassociated* initiator molecules are active in initiation. Hence, if K_e is small, the rate of initiation will be proportional to $[(\mathrm{RLi})_n]^{1/n}$, where the latter represents virtually the *total* initiator concentration.

Although this simple concept may be superficially attractive, it does not really merit serious consideration as a scientific hypothesis on several grounds, as follows:

(i) The experimental methods used to determine the states of association shown in Table 3.2 required *high concentrations* of the organolithium compounds, compared with the much lower values used in studies of polymerization kinetics. Hence the state of association of an initiator under *polymerization conditions* cannot be assumed on this basis.

(ii) An examination of Tables 3.1 and 3.2 clearly shows that there is really no reliable correspondence between states of association and reaction order of organolithium initiators, especially since the latter appears to be influenced by type of monomer, solvent, and concentration.

(iii) The mechanism proposed in Eqs. (3.15) and (3.16) fails to take into account the strong cross-association which has been found to occur between the organolithium growing chains and the initiator, and which is discussed in a later part of this book. Such cross-association can be expected to have a marked effect on the initiation reaction.

(iv) The dissociation equilibrium shown in Eq. (3.15) has been challenged by Brown,[43,44] who pointed out that, if $n = 6$, as proposed in the case of *n*-butyllithium, such a dissociation would involve an enthalpy change of about 100 kcal, much too high to account for any reasonable concentration of the supposedly active unassociated species.

Based on his objections to the reaction scheme expressed by Eqs. (3.15) and (3.16), Brown[43,44] suggested instead that it would be more plausible to assume a direct reaction between monomer and initiator in its associated form, and that the fractional order of this reaction could be accounted for by several possible mechanisms. Several others[45-49] have subsequently supported this view and have demonstrated various ways in which this reaction between an associated initiator and the monomer could exhibit a fractional reaction order.

It would appear obvious, therefore, that the initiation mechanism proposed in Eqs. (3.15) and (3.16) is much too simplistic to merit acceptance. Thus the phenomenon described under (iii) above, i.e., the cross-association between initiator and growing chains, would be sufficient to render such a mechanism inoperative. In such a situation, as the initiator concentration is depleted and the chain-end concentration increases, the remaining initiator would be *mainly present* in the cross-associated complex. Without a knowledge of the reactivity of the initiator bound in the cross complex, it is impossible to carry out a kinetic analysis. It is intuitively obvious, however, that the low order of dependence of initiation rate on concentration could very well be accounted for by such a cross-complexation.

The effect of monomer and initiator concentration on the reaction order can also lead to misleading results, as suggested under item (i) above. Thus Worsfold and Bywater[18] were the first to propose the reaction scheme defined by Eqs. (3.15) and (3.16) in the initiation of styrene polymerization by *n*-butyllithium in benzene at 30°C, in which they found a reaction order of $\frac{1}{6}$. However, in that work they used a very low ratio of styrene to initiator, about 10, i.e., much lower than normally used to prepare a high polymer. Under these conditions, their initiation rates were so slow, relative to propagation, that they postulated a minimum molecular weight of 7×10^3 as necessary to ensure complete consumption of the *n*-butyllithium. However, more recent work[26a,50] has clearly demonstrated that this prediction was not true, and that the initiation rates were *much faster* when the initial monomer–initiator molar ratio was high enough to produce the molecular weights indicated above, i.e., about 100/1 instead of 10/1. This is shown by the distribution curve in Fig. 3.1, which offers convincing evidence that the initiation reaction must be sufficiently rapid under these conditions to produce such a narrow distribution curve for an average molecular weight (M_n) of only 11,700. Analysis after completion of the polymerization showed no residual *n*-butyllithium, as would be expected from Fig. 3.1, and the good agreement between the predicted stoichiometric molecular weight (M_s) and the osmometric value (M_n) also confirms complete utilization of initiator.

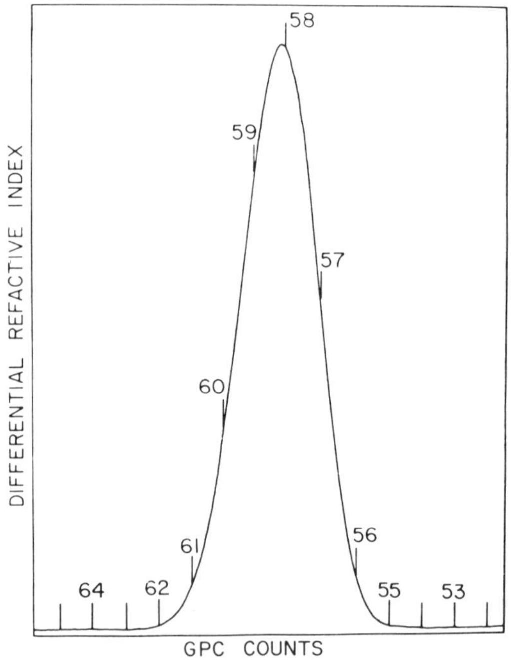

Fig. 3.1 Gel permeation chromatogram of polystyrene prepared with *n*-butyllithium in benzene at 30°C. $M_s = 1.15 \times 10^4$; $M_n = 1.17 \times 10^4$; $M_w = 1.26 \times 10^4$. Initial monomer concentration 0.1 *M*.[26a, 50]

It is obvious from the above results that the initiation reaction between alkyllithium and vinyl monomers in hydrocarbon media is a complex one, and is sensitive to concentrations of all the reactants. It probably involves a reaction between the monomer and the *associated* form of the initiator, both with itself and cross-complexed with the growing chains. The above considerations also amply demonstrate the folly of deducing a state of association from the observed kinetic order of a reaction, without actual direct evidence for such an association *under the conditions of the reaction.*

Despite the complexities of these initiation reactions, it has been possible to learn enough about them to compare the reactivities of the initiators toward styrene and the dienes, and the effect of various hydrocarbon solvents. Thus, in *decreasing* order of reactivity, the effects of the organic moiety (R group) of the initiator are as follows:

Styrene: menthyl > *s*-butyl > *i*-propyl > *n*-butyl and ethyl > *t*-butyl

Dienes: menthyl > *s*-butyl > *i*-propyl > *t*-butyl > *n*-butyl and ethyl

and the effect of solvents:

Solvent: toluene > benzene > *n*-hexane > cyclohexane

One of the interesting features of the above series is the fact that the initiators show the same relative reactivities toward styrene and the dienes (e.g., butadiene and isoprene), except in the case of *t*-butyllithium, which is a moderately fast initiator for the dienes but a very slow one for styrene. This low reactivity of *t*-butyllithium toward styrene had been known for some time and had been ascribed to the steric hindrance between the two bulky substituent groups involved, i.e., *t*-butyl and phenyl. However, recent work[51,52] has shown that there is apparently a rapid *initial* reaction between styrene and *t*-butyllithium for a brief period, followed by a rapid decrease in rate to a long-drawn-out initiation reaction, which leads to the formation of a low molecular weight "tail" in the polymer. Figure 3.2 shows that such a polymer does indeed possess a long low-molecular-weight tail without a corresponding high-molecular-weight tail, which would be the case for "normal" slow initiation.

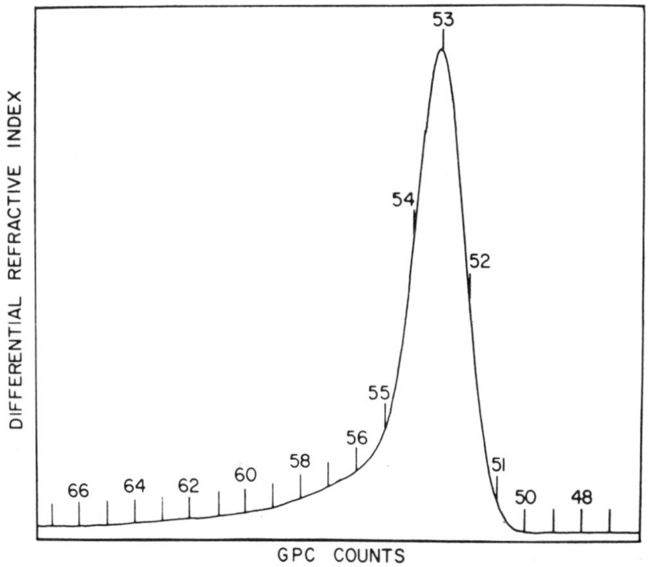

Fig. 3.2 Gel permeation chromatograph of polystyrene prepared with *t*-butyllithium in benzene at 30°C. $M_s = 1.9 \times 10^4$; $M_n = 2.9 \times 10^4$; $M_w = 3.8 \times 10^4$. Initial monomer concentration 0.5 *M*. Unreacted initiator was present at end of polymerization.[26a,52]

This asymmetric distribution curve seems to confirm the observed kinetics, i.e., that only a portion of the initiator reacts rapidly, while the remainder is much less reactive.

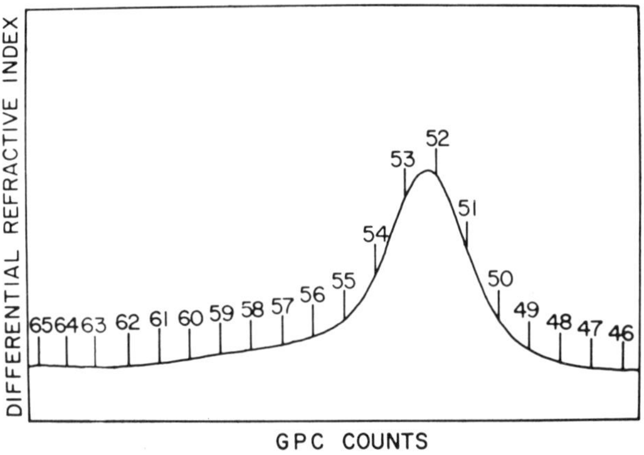

Fig. 3.3 Gel permeation chromatogram of polystyrene prepared with t-butyllithium in cyclohexane at 35°C. $M_s = 1.11 \times 10^4$; $M_n = 4.0 \times 10^4$; $M_w = 5.2 \times 10^4$. Initial monomer concentration 0.5 M.[26a, 52]

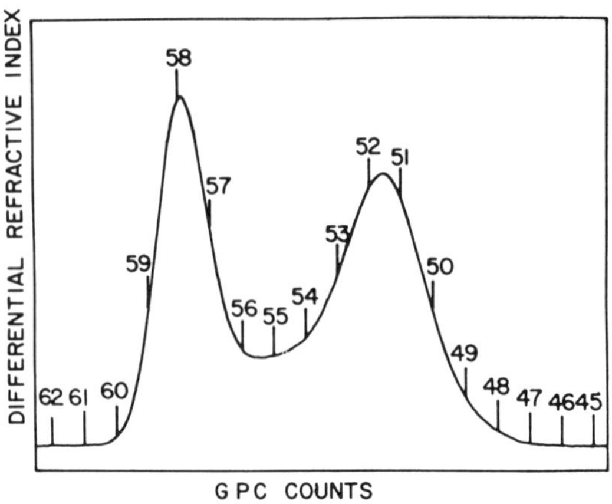

Fig. 3.4 Gel permeation chromatogram of polystyrene prepared by a second monomer addition to polystyrene of Fig. 3.3. $M_s = 2.22 \times 10^4$; $M_n = 2.3 \times 10^4$, $M_w = 4.1 \times 10^4$. Monomer concentration after second monomer addition 0.5 M.[26a, 52]

This phenomenon, which occurs in the t-butyllithium-initiated polymerization of styrene, is further elucidated by an examination of Figs. 3.3 and 3.4, which represent, respectively, the molecular weight distribution of a polystyrene formed in cyclohexane at 35°C (Fig. 3.3) and that formed after addition of a second increment of polystyrene together with 1 vol % H_4-furan to the same mixture (Fig. 3.4). It is obvious that the second peak (count 58) in the bimodal distribution curve of Fig. 3.4 resulted from initiation by the residual *t*-butyllithium, which was apparently inactive or "dormant" prior to the second addition of styrene. This inactive form of the initiator, representing about 75% of the original, was apparently activated by the addition of the H_4-furan with the second increment of styrene, so that all of the initiator was thus utilized. This is strikingly confirmed by the fact that the original polymer (Fig. 3.3) showed a much higher observed molecular weight (M_n) than predicted (M_s), while the two molecular weight averages show excellent agreement for the final polymer (Fig. 3.4).

The behavior of *t*-butyllithium with styrene presents an excellent example of the complexities that may mark the kinetics of initiation reactions. One may speculate that, once initiation starts in this case, the newly formed polystyryllithium chain ends form strong cross-complexes with the residual initiator, and that these cross-complexes show little, if any, reactivity toward styrene. (As a matter of fact, the 75% of residual initiator found in this system after its initial rapid consumption corresponds remarkably well with the 3:1 ratio of initiator to chain ends actually determined experimentally for ethyllithium in isoprene polymerization, as will be discussed in a later section.) In any event, there is convincing evidence that the initiation reaction between organolithium initiators and vinyl monomers in hydrocarbon media is not as simple and straightforward as represented by Eqs. (3.15) and (3.16).

b. Polar Vinyl Monomers

Much less is known about the initiation reaction for monomers other than styrene and the dienes. Most of the vinyl monomers having polar substituents participate, as might be expected, in various side reactions with the organolithium initiators, so that these cannot be classified as nonterminating (living) systems. (Even though there may be little or no direct side reactions between the monomer and the initiator, such termination steps may occur during the propagation of the chain, again leading to nonliving systems.) The reader is referred to Table 2.1 for listings and literature references to work done with this class of monomers.

The one group of polar vinyl monomers which have been most extensively investigated are the acrylates, specifically methyl methacrylate. In this case it has been shown[53,54] that organolithium compounds, besides initiating polymerization through the vinyl group, also react substantially with the

carbonyl function; thus

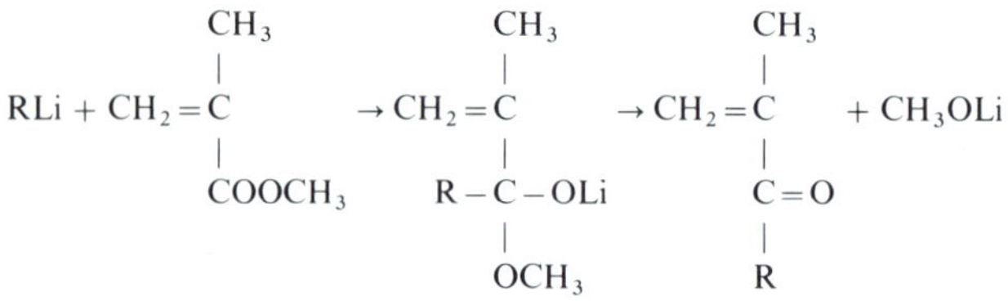

(3.17)

This Grignard-type reaction (which apparently occurs also with organomagnesium compounds) generates methanol upon hydrolysis. The competition between this reaction and the initiation of polymerization varies with type of organolithium compound, solvent, and temperature. Thus, with *n*-butyllithium in toluene at $-30°C$, about 60% of the initiator disappears by this side reaction, while 1,1-diphenylhexylhexyllithium (the adduct of *n*-butyllithium and 1,1-diphenylethane), undergoes none of this side reaction. However, the growing chains initiated by the latter initiator undergo a similar slow reaction during the course of the polymerization, so that about 15–20% of the chains undergo termination in this way. On the other hand, in ether solvents, at low temperatures, this side reaction can be greatly reduced.

c. Heterocyclics

Although it had been shown long ago that bases, such as the sodium and potassium alkoxides, can initiate polymerization of ethylene oxide by ring opening, it was found that lithium alkoxides were poor initiators, presumably because of their lower basicity. Lithium carbanions, on the other hand, react readily with ethylene oxide to form the alkoxide by ring opening, but any subsequent chain growth reaction is very slow. Metallation can also occur readily, especially at higher temperatures.

Propylene oxide is relatively inert to alkali metal initiators including organolithium, presumably due to the electropositive character of the methyl substituent, and higher oxides are thermodynamically insuitable for the ring opening step.[55, 56]

Organolithium initiators are effective in initiating polymerization of cyclic sulfides, but the mechanism differs sharply from that of ethylene oxide. Thus it has been shown[57] that the reaction between alkyllithium and propylene sulfide proceeds by the following mechanism:

(3.18)

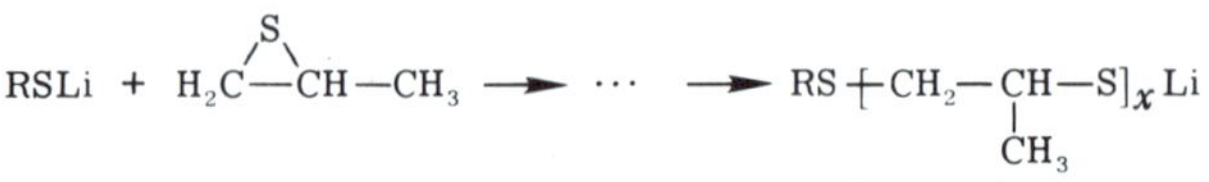

(3.19)

Stoichiometric formation of propylene by reaction (3.16) was confirmed; also, polymerization of this monomer was actually initiated by lithium thiolates.[57] A similar initiation mechanism was found for ethylene sulfide, with formation of ethylene, but this was complicated by the noticeable occurrence of metallation of the monomer by the alkyllithium. This did not, however, interfere with the subsequent polymerization of the ethylene sulfide.

The difference in behavior between propylene oxide and propylene sulfide is carried over into the next higher cyclic, i.e., thiacyclobutane. Whereas the four-membered cyclic oxide is inert in this respect, the sulfur analog reacts as follows[57]:

$$\begin{array}{l} RLi + CH_2-S \\ \qquad\;\; | \qquad | \\ \qquad CH_2-CH_2 \end{array} \rightarrow R-S-CH_2-CH_2-CH_2Li \tag{3.20}$$

This initiation step is followed by the expected propagation step, leading to the usual nonterminating anionic polymerization. The fact that the growing chain in this case is *carbanionic* in nature is confirmed by the fact that it can initiate a block copolymerization of styrene.[57,58] Furthermore, lithium thiolates were found[57] to be ineffective in initiating the polymerization of these thietanes. Thiacyclopentane and higher-membered rings were inert to any of these initiators.

It is interesting to note that both Eqs. (3.18) and (3.20) indicate a nucleophilic attack on the divalent sulfur, but the subsequent pathways are different. In the case of the 3-membered ring, the elimination of propylene is the lowest energy path for conversion of the highly strained ring, while for the 4-membered ring no such pathway is available, since formation of cyclopropane would involve even more energy, so that direct ring opening ensues with consequent formation of a growing carbanion.

Another heterocyclic which is susceptible to ring opening polymerization initiated by organolithium is the cyclic siloxane trimer hexamethylcyclotrisiloxane. Unlike the more common cyclic tetramer, which requires strong bases, e.g., KOH, for the ring-opening polymerization,[59,60] the highly strained trimer can be polymerized by organolithium initiators, and therefore has the advantage that the weakly basic lithium silanolate chain end is not active in bond rearrangement with the siloxane bonds of the linear chains. The net result is that the growing lithium silanolate chain is truly a living polymer, and can lead to the usual narrow molecular weight distribution.[61] It is also possible in this way to make block copolymers of vinyl monomers and siloxanes having a well-defined narrow distribution of chain lengths and blocks.[62-64]

References

1. W. Schlenk, J. Appenrodt, A. Michael, and A. Thal, *Ber. Dtsch. Chem. Ces.* **47**, 479 (1914).
2. W. Schlenk and E. Bergmann, *Justus Liebigs Ann. Chem.* **463**, 91 (1928).

3. D. Lipkin, D. E. Paul, J. Townsend, and S. I. Weissman, *Science* **117**, 534 (1953); N. D. Scott, U. S. Patent 2,181,777 (1939).
4. D. E. Paul, D. Lipkin, and S. I. Weissman, *J. Am. Chem. Soc.* **78**, 116 (1956).
5. N. D. Scott, U. S. Patent 2,181,771 (1939).
5a. N. D. Scott, J. F. Walker, and V. L. Hansley, *J. Am. Chem. Soc.* **58**, 2442 (1936); J. F. Walker and N. D. Scott, *ibid.* **60**, 951 (1938).
6. G. S. Whitby and H. L. Stephens, Report CR-3133 (1952), NTIS, *U.S. Dep. Commer., Off. Tech. Serv., PB Rep.* **PB 118310** (1955).
7. M. Szwarc, M. Levy, and R. Milkovich, *J. Am. Chem. Soc.* **78**, 2656 (1956).
8. M. Morton, A. Rembaum, and E. E. Bostick, *J. Polym. Sci.* **32**, 530 (1958).
9. D. H. Richards and M. Szwarc, *Trans. Faraday Soc.* **55**, 1644 (1959).
10. T. J. Dudek, Ph.D. Dissertation, University of Akron, Akron, Ohio, 1961.
11. S. Boileau, G. Champetier, and P. Sigwalt, *Makromol. Chem.* **69**, 180 (1963).
11a. S. Boileau, G. Champetier, and P. Sigwalt, *J. Polym. Sci., Part C* **16**, 3021 (1967).
12. K. Ziegler, L. Jakob, H. Willthan, and A. Wenz, *Justus Liebigs Ann. Chem.* **511**, 64 (1934).
13. R. E. Robertson and L. Marion, *Can. J. Res., Sect. B* **26**, 657 (1948).
14. C. G. Overberger and N. Yamamoto, *J. Polym. Sci., Polym. Chem. Ed.* **4**, 3101 (1966).
15. R. K. Graham, D. L. Dunkelberger, and W. E. Goode, *J. Am. Chem. Soc.* **82**, 400 (1960).
16. A. A. Korotkov, S. P. Mitzengendler, and J. L. Danzig, *J. Polym. Sci., Part B* **4**, 809 (1966).
17. K. F. O'Driscoll, E. N. Ricchezza, and J. E. Clark, *J. Polym. Sci., Part A* **3**, 3241 (1965).
18. D. J. Worsfold and S. Bywater, *Can. J. Chem.* **38**, 1891 (1960).
19. A. F. Johnson and D. J. Worsfold, *J. Polym. Sci., Part A* **3**, 449 (1965).
20. H. Hsieh, *J. Polym. Sci., Part A* **3**, 163 (1965).
21. G. M. Burnett and R. N. Young, *Eur. Polym. J.* **2**, 329 (1966).
22. D. J. Worsfold and S. Bywater, *Can. J. Chem.* **42**, 2884 (1964).
23. S. Bywater and D. J. Worsfold, *J. Organomet. Chem.* **10**, 1 (1967).
24. V. Zgonnik, E. Schadrina, K. Kalnin'sh, and B. Erussalimsky, *Makromol. Chem.* **174**, 81 (1973).
25. A. Guyot and J. Vialle, *J. Macromol. Sci., Chem.* **A4**, 79 (1970); *J. Polym. Sci., Part B* **6**, 403 (1968).
26. J. E. L. Roovers and S. Bywater, *Macromolecules* **1**, 328 (1968).
26a. M. Morton and L. J. Fetters, *Rubber Chem. Technol.* **48**, 359 (1975).
27. F. Hein and H. Schram, *Z. Phys. Chem., Abt. A* **151**, 234 (1930).
28. T. L. Brown and M. T. Rogers, *J. Am. Chem. Soc.* **79**, 1859 (1957).
29. T. L. Brown, D. W. Dickerhoff, and D. A. Baffus, *J. Am. Chem. Soc.* **84**, 1371 (1962).
30. T. L. Brown, R. L. Gerteis, D. A. Baffus, and J. A. Ladd, *J. Am. Chem. Soc.* **86**, 2134 (1964).
31. T. L. Brown, J. A. Ladd, and G. N. Newman, *J. Organomet. Chem.* **3**, 1 (1965).
32. T. L. Brown, *Acc. Chem. Res.* **1**, 23 (1968).
33. H. L. Lewis and T. L. Brown, *J. Am. Chem. Soc.* **92**, 4664 (1970).
34. D. Margerison and J. P. Newport, *Trans. Faraday Soc.* **59**, 2058 (1963).
35. G. Wittig, F. J. Meyer, and G. Lange, *Justus Liebigs Ann. Chem.* **571**, 167 (1951).
36. M. Weiner, C. Vogel, and R. West, *Inorg. Chem.* **1**, 654 (1962).
37. D. Margerison and J. D. Pont, *Trans. Faraday Soc.* **67**, 353 (1971).
38. R. H. Baney and R. J. Krager, *Inorg. Chem.* **3**, 1967 (1964).
39. G. E. Hartwell and T. L. Brown, *Inorg. Chem.* **3**, 1656 (1964).
40. J. B. Smart, R. Hogan, P. A. Scherr, M. T. Emerson and J. P. Oliver, *J. Organomet. Chem.* **64**, 1 (1974).
41. W. H. Glaze, J. E. Hanicak, M. L. Moore, and J. Chaudhuri, *J. Organomet. Chem.* **44**, 39 (1972).
42. W. H. Glaze and C. H. Freman, *J. Am. Chem. Soc.* **91**, 7198 (1969).
43. T. L. Brown, *J. Organomet. Chem.* **5**, 191 (1966).

44. T. L. Brown, *Adv. Organomet. Chem.* **3**, 365 (1965); *Ann. N. Y. Acad. Sci.* **136**, 95 (1966); *Pure Appl. Chem.* **23**, 447 (1970).
45. W. Glaze, J. E. Hanicak, D. J. Berry, and D. P. Duncan, *J. Organomet. Chem.* **44**, 49 (1972).
46. P. D. Bartlett, C. V. Goebel, and W. P. Weber, *J. Am. Chem. Soc.* **91**, 7425 (1969).
47. W. H. Glaze, J. Lin, and E. G. Felton, *J. Org. Chem.* **30**, 248 (1966).
48. J. P. Oliver, J. B. Smart, and M. T. Emerson, *J. Am. Chem. Soc.* **88**, 4101 (1966).
49. H. S. Makowski and M. Lynn, *J. Macromol. Chem.* **1**, 443 (1966).
50. L. J. Fetters and H. Asai, private communication.
51. L. J. Fetters and M. Morton, *Macromolecules* **7**, 552 (1974).
52. L. J. Fetters and J. Rupert, private communication.
53. W. E. Goode, F. H. Owens, and W. L. Myers, *J. Polym. Sci.* **47**, 75 (1960).
54. F. H. Owens, W. L. Myers, and F. E. Zimmerman, *J. Org. Chem.* **26**, 2288 (1961).
55. G. Gee, W. C. E. Higginson, P. Levsley, and K. J. Taylor, *J. Chem. Soc.* p. 1338 (1959).
56. G. Gee, W. C. E. Higginson, and G. T. Merrall, *J. Chem. Soc.* p. 1345 (1959).
57. M. Morton and R. F. Kammereck, *J. Am. Chem. Soc.* **92**, 3217 (1970).
58. M. Morton, R. F. Kammereck, and L. J. Fetters, *Macromolecules* **4**, 11 (1971); *Br. Polym. J.* **3**, 120 (1971).
59. W. T. Grubb and R. C. Osthoff, *J. Am. Chem. Soc.* **77**, 1405 (1955).
60. M. Morton and E. E. Bostick, *J. Polym. Sci., Part A* **2**, 523 (1964).
61. J. G. Saam, D. J. Gordon, and S. Lindsey, *Macromolecules* **3**, 1 (1970).
62. W. G. Davies and D. P. Jones, *Polym. Prepr., Am. Chem. Soc., Div. Polym. Chem.* **11**, 447 (1970); "Colloidal and Morphological Behavior of Block and Graft Copolymers." Plenum, New York, 1971.
63. A. Ward, T. Kendrick, and J. G. Saam, *Polym. Prepr., Am. Chem. Soc., Div. Polym. Chem.* **15** (1), 183 (1974); *in* "Copolymers, Polyblends, and Composites" (N. A. J. Platzer, ed.), p. 300. American Chemical Society, Washington, D.C. 1975.
64. M. Morton, Y. Kesten, and L. J. Fetters, *Polym. Prepr., Am. Chem. Soc., Div. Polym. Chem.* **15** (2), 175 (1974); *Appl. Polym. Symp.* **26**, 113 (1975).

Chapter

4 The Propagation Reaction

In the study of polymerization reactions, the single most important step to consider is, of course, the chain growth or propagating reaction. One special feature of homogeneous anionic polymerization is that, under proper conditions, this step may be studied independently of any interference from either the termination or initiation processes. In these circumstances the propagation reaction, which is homogeneous in nature, may be expected to have the following characteristics:

(i) It should be independent of the mode of initiation, whether by alkali metals, aromatic complexes, or organolithium.
(ii) It would be expected to depend on the nature of the counterion introduced by the initiator, which may be associated with the growing anionic chain end.
(iii) It would be expected to be affected by the nature of the medium (solvent), which could influence the character of the growing chain end.

Regardless of the above considerations, the propagation step should, in principle, be a simple bimolecular reaction between the growing chain and the monomer, as it is, for example, in free radical polymerization. Since, as indicated above, many of the anionic systems can be kept termination-free, it has been possible from the start to study the propagation reaction independently. These studies led to the rather disappointing results that the chain growth process, in the majority of cases, does not obey the kinetic rules of a simple bimolecular reaction, indicating instead a more complex mechanism. There appears to be general agreement that the propagation rate is always first order in monomer, but rarely first order with respect to the concentration of growing chains (i.e., initiator concentration). As in the case of the initiation reaction, the kinetics of propagation exhibit various fractional orders relative to initiator concentration, and these depend on the two controlling factors, viz., counterion and solvent. These two parameters must therefore be taken into consideration in any study of polymerization by anionic initiators. The effect of solvents can be especially dramatic, polar and nonpolar media having markedly different effects on polymerization rates.

1. Vinyl Monomers

The most thoroughly investigated monomers used in homogeneous anionic polymerization have been styrene (and some of its derivatives) and the 1,3-dienes. This is not really surprising in view of the importance of these monomers in technology, and also because of the variety of isomeric chain structures that result in the case of the dienes. The effect of solvents and counterions on the kinetics of these reactions is immediately obvious. Thus the polymerization of styrene by sodium naphthalene, which occurs in an ether solvent, such as H_4-furan, is almost instantaneous even at $-70°C$, and can

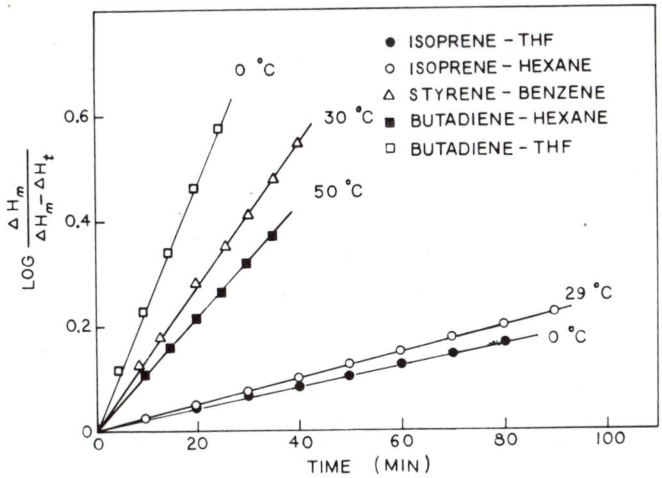

Fig. 4.1 Propagation rates of styrene, isoprene, and butadiene, initiated by *n*-butyllithium in various solvents, plotted as first-order kinetics.[1-3] [RLi] = 1.65×10^{-3} mole liter^{-1}. (Courtesy of John Wiley and Sons.)

only be carried out (in the laboratory) by dropwise addition of the monomer to the initiator solution. On the other hand, with organolithium initiators and nonpolar solvents such as benzene, the rate of polymerization is measured in hours. Some typical first-order rate plots for styrene, butadiene, and isoprene in organolithium systems[1-3] are shown in Fig. 4.1, and illustrate clearly the marked effect of solvent type on rate. The effect of these solvents on the chain microstructure of the polydienes is discussed in a later section.

As stated earlier, although the propagation rates invariably obey first-order kinetics with respect to the monomer, they generally do not exhibit a similar rate dependence on initiator (i.e., growing chain) concentration, the kinetic order usually being fractional. Here again, the solvent and counterion exert a marked influence, as does the type of monomer, and a special distinction must therefore be made between polar and nonpolar solvents.

a. Polar Solvents

The kinetics of the propagation step in a polar solvent such as H_4-furan is shown in Fig. 4.2 for butadiene and isoprene, initiated by *n*-butyllithium.[1-3] Figure 4.2 shows the expected first-order rate dependence on initiator concentration, indicating that the growing chain is a single species, with a propagation rate constant of 4.5 liter mole^{-1} sec^{-1} for butadiene and 0.36 liter mole^{-1} sec^{-1} for isoprene at 60°C.

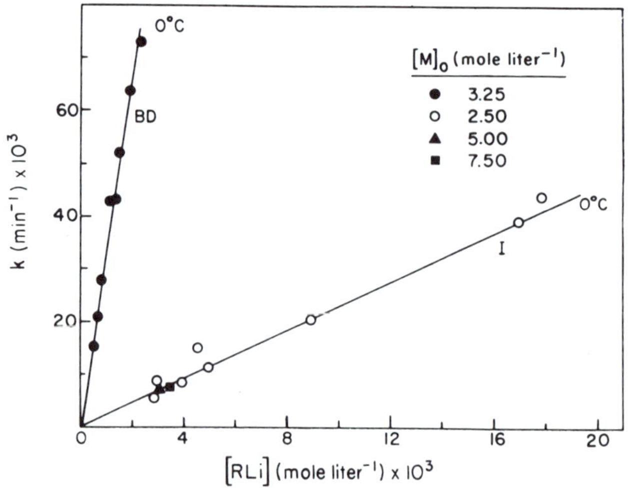

Fig. 4.2 Dependence of rates of propagation of butadiene (BD) and isoprene (I) on chain-end concentration (initiator concentration) in H_4-furan at 0°C (k = first-order rate constant).[1-3] (Courtesy of John Wiley and Sons.)

Similar results for these dienes were obtained by others with organolithium systems in polar solvents such as H_4-furan or amines.[4,5] When the counterion was sodium from sodium naphthalene initiator, and the monomer was styrene, the propagation rate in dioxane solution was also found[6,7] to be first order with respect to both monomer and initiator, the value of k_p, the propagation rate constant, being about 4 liter mole^{-1} sec^{-1} at 25°C.

On the other hand, Fig. 4.3 clearly indicates that this was not the case for styrene with sodium as counterion in H_4-furan.[8] The *increase* in the first-order rate dependence on initiator concentration with *dilution*, shown in Fig. 4.3, strongly suggested the activity of a dissociated form of the growing chain end.

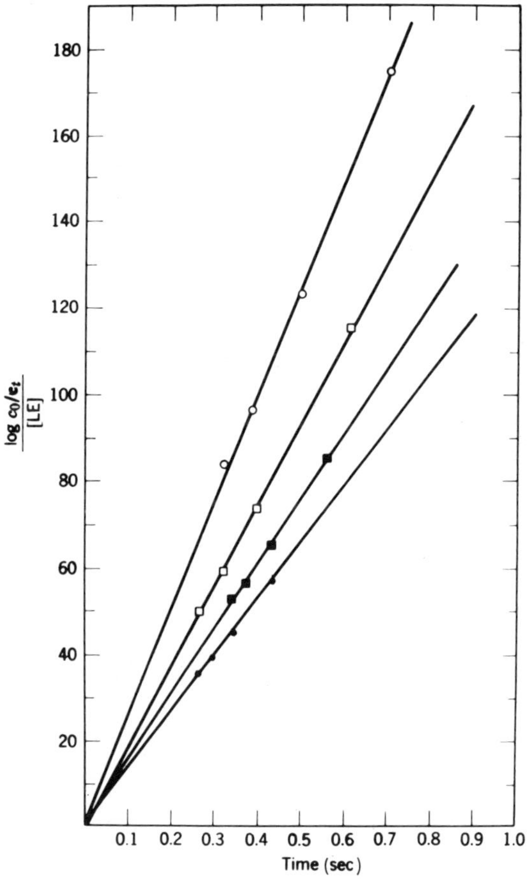

Fig. 4.3 Effect of chain-end concentration ([LE]) on first-order propagation rates of styrene in H_4-furan at 25°C, initiated by sodium naphthalene. [LE] = 1.1×10^{-3} (○), 7.6×10^{-3} (□), 7.4×10^{-3} (−7°C) (■), 14.5×10^{-3} (●).[8] (Courtesy of John Wiley and Sons.)

In polar solvents such a dissociation of an organoalkali chain end can be expected to be ionic in nature.[9]

On this basis, it was suggested[10-12] that the active species during the styrene propagation reaction in polar solvents could be dual in nature, consisting of an equilibrium mixture of ion pairs and ions; thus

$$\begin{array}{ccc} RS_x^-Na^+ & \overset{K_e}{\rightleftharpoons} & RS_x^- + Na^+ \\ S\downarrow k_p^{\pm} & & S\downarrow k_p^- \\ RS_{x+1}^-Na^+ & & RS_{x+1}^- + Na^+ \end{array} \tag{4.1}$$

where $RS_x^-Na^+$ represents the growing polystyrene chain having an ion-pair active end, RS_x^- represents the growing polystyrene chain with a free anion active end, K_e is the ionization constant, $k_p^{\pm}$ and k_p^- are the propagation rate constants for ion pairs and free ions, respectively, and S is styrene monomer. In such a system, the overall propagation rate R_p can be expressed as

$$R_p = k_p^{\pm}[RS_x^-Na^+][S] + k_p^-[RS_x^-][S] \tag{4.2}$$

Taking into account the dissociation equilibrium shown in Eq. (4.1), we have

$$R_p = k_p^{\pm}[RS_x^-Na^+][S] + k_p^-K_e^{1/2}[RS_x^-Na^+]^{1/2}[S] \tag{4.3}$$

or

$$R_p/[S][RS_x^-Na^+]^{1/2} = k_p^-K_e^{1/2} + k_p^{\pm}[RS_x^-Na^+]^{1/2} \tag{4.4}$$

If ion-pair dissociation is low, i.e., K_e is small, then the approximate Eq. (4.5) can be written

$$R_p[S]^{-1}[RS_xNa]_t^{-1/2} = k_p^-K_e^{1/2} + k_p^{\pm}[RS_xNa]_t^{1/2} \tag{4.5}$$

where $[RS_xNa]_t$ represents the *total* concentration of growing chains, both ion pair and free ion, i.e., the initiator concentration.

Equation (4.5) makes it possible to solve for the values of $k_p^{\pm}$ and $k_p^-K_e^{1/2}$ from the linear plot of $R_p[S]^{-1}[RS_xNa]^{-1/2}$ versus $[RS_xNa]^{1/2}$. This means that the ion-pair propagation rate constant $k_p^{\pm}$ can be obtained directly, while the free-ion propagation rate constant k_p^- may be calculated if K_e is known. The latter is actually measurable by the usual conductimetric techniques, since the degree of ionization is sufficiently large ($K_e \sim 10^{-7}$–10^{-10}) in these solvents. Hence the absolute values of k_p^- have also been determined in these cases.

As might be expected, the most thoroughly investigated monomer in this connection has been styrene, and values for K_e, k_p^-, and $k_p^{\pm}$ have been obtained for a variety of solvents and alkali counter-ions. In these systems, the values of K_e generally range from 10^{-12} to 10^{-6} at room temperature,[13-16] depending on solvent and counterion, the former apparently being much more important than the latter. The ionization is apparently governed not only by the dielectric constant of the solvent, as usual, but also by the ability of the solvent to interact specifically with the ion pair, presumably by "solvating" (coordinating with) the cation. Thus solutions of polystyrylsodium in dioxane ($D = 2.4$) showed no measurable conductivity[17] and hence presumably no ionization. In H_4-furan ($D = 8.2$) the same system[13] showed a K_e of 1.5×10^{-7}, but in dimethoxyethane ($D = 8.0$), the K_e increased[14] to 1.4×10^{-6}, a change which can only be attributed to the much greater solvating power of the latter solvent.

As for k_p^-, the propagation rate constant for the free polystyryl anion, experimental values show a range of 3×10^4 to 7×10^4 liter mole^{-1} sec^{-1} at 25°C, and these are apparently independent of solvent type,[18] as expected. This is not the case for $k_p^{\pm}$, the ion-pair propagation rate constant, which shows a strong dependence on counterion and solvent. However, this dependence indicates some interesting anomalies, as shown in Table 4.1.

TABLE 4.1 Ion-Pair Propagation Rate Constants[a] in Anionic Polymerization of Styrene

Solvent	Li^+	Na^+	K^+	Rb^+	Cs^+	References
Dioxane	0.94	3.4	20	21.5	24.6	17
H_4-pyran	< 10; 19.5[b]	14; 17[b]	73; 30[b]	83	53	19, 20; 15
H_4-furan	160	80	~60	~50	22	12
Dimethoxyethane	—	3600	—	—	—	14

[a] $k_p^{\pm}$, liter mole^{-1} sec^{-1} at 25°C.
[b] At 30°C.

Although the data are collected from different sources and show considerable discrepancies, certain trends are evident. Thus, in a solvent of low solvating power, such as dioxane, the ion-pair activity increases from the least electropositive lithium counterion to the most electropositive cesium counterion, as would be expected. However, the reverse order of activity is noted for a highly solvating medium such as H_4-furan. This has been ascribed to the formation of the more reactive "solvent-separated" ion pairs in such solvents, and these would be formed more readily in the case of the more covalently bonded, smaller lithium cation than for the larger, more electropositive cations. The effect of the most highly solvating dimethoxyethane is dramatically illustrated in the one rate that was determined for the sodium counterion. Hence it appears that there can be more than one type of propagating ion pair in these polymerizations.

It might be appropriate at this point to take note of the basic nature of these propagating polymer chains which may assume different characters, i.e., free ions, "contact" ion pairs, solvent-separated ion pairs, etc. These various identities will, of course, be assumed by each of the growing chains, involving ionic dissociation and solvation equilibria. Yet the occurrence of such dynamic equilibria during the growth of the chain should not have any effect on the molecular weight distribution because the *rates* of dissociation, solvation, etc., are generally some *orders of magnitude* greater than the *rates of chain propagation*. Thus, although these processes follow random statistics, they are too rapid to disturb the Poisson distribution of molecular weights resulting from the living character of the propagation reaction.

Unfortunately, very little additional information is available about the behavior of other vinyl monomers in these polar solvents. The kinetics of butadiene and isoprene polymerization with organolithium initiators[1-3] has already been mentioned previously, as indicating the presence of a single active species, presumably an ion pair, as the principal participant. The activity of the sodium–polybutadiene ion pair has also been determined[21] in H_4-furan, and appears to be greater than that of the lithium counterpart in the same system,[2] perhaps indicating little solvation effect on polydienyl ion pairs. Work of this type has apparently also been done on other monomers such as the methylstyrenes,[22-28] methoxystyrenes,[29] and vinylpyridine,[30,31] but it may be difficult to draw any conclusions in these cases in view of the strong possibility of side reactions during polymerization. The reader is therefore referred to Table 2.1 for direct access to the pertinent literature.

b. Nonpolar Solvents

In the case of nonpolar media, i.e., hydrocarbon solvents, kinetic data are available only for styrene and the dienes, viz., butadiene and isoprene. Here again the rate dependence on monomer is generally first order (although higher orders have been claimed in some cases), but its dependence on active centers varies from first order to fractional orders, as the counterion, and even the solvent, is changed. Thus in the homogeneous polymerization of styrene, it has been reported that the reaction order changes from first in the case of cesium initiators[32] to one-half for lithium[1-3,33] and sodium[32] initiators, with potassium varying from one-half at higher concentrations to first at lower concentrations.[32] The first-order dependencies seem to indicate a direct reaction between chain ends and monomer, but anything less has been taken as indicating the presence of a dissociation phenomenon. Since ionic dissociation is not an acceptable hypothesis in such nonpolar media, such experimental results have been presumed to point to another type of association–dissociation equilibrium.

Furthermore, in the case of the dienes such as isoprene and butadiene, organolithium initiation leads to propagation kinetics showing an even lower fractional order relative to growing chains, depending on which solvent is used. Unfortunately, the literature on this topic is even more contradictory and confusing than for the case of styrene. A compendium of data on the kinetics of

TABLE 4.2 Propagation Rates in Organolithium Polymerization in Nonpolar Solvents

Solvent	Initiator	[Initiator]	Order of monomer dependence	Order of initiator dependence	References
		Styrene			
Benzene	C_2H_5Li	$(1–50) \times 10^{-4}$	1	0.5	3
	n-C_4H_9Li	$10^{-3}–10^{-1}$	2	0	34
		$< 2 \times 10^{-2}$	1	1	35
		$> 2 \times 10^{-2}$	Variable	0	35
		$(8–70) \times 10^{-3}$	1	0.5	36
		$1.6 \times 10^{-5}–3.9 \times 10^{-2}$	1	0.5	33
Toluene	s-C_4H_9Li	$(3–40) \times 10^{-4}$	1	0.5	37
Cyclohexane	n-C_4H_9Li	$(5–100) \times 10^{-5}$	1	0.5	38
	s-C_4H_9Li	$(7–70) \times 10^{-4}$	1	0.5	37
		Isoprene			
n-Hexane	n-C_4H_9Li	$(2.4–57) \times 10^{-4}$	1	0.5	1–3
		$(2–140) \times 10^{-5}$	1	0.5	39
		$(2–16) \times 10^{-3}$	1–2	0	40
		$(5–100) \times 10^{-4}$	1	0.25	41
	t-C_4H_9Li	$(1–10) \times 10^{-3}$	1	0.5	42
n-Heptane	n-C_4H_9Li	$10^{-6}–10^{-2}$	1	0.17–0.5	43
		$< 5 \times 10^{-3}$	1–2	0.25	44
		$(5–20) \times 10^{-3}$	1–2	0–0.25	44
		$> 20 \times 10^{-3}$	1–2	Negative	44
Cyclohexane	C_2H_5Li	$(1–100) \times 10^{-4}$	1	0.25	41
		$(0.6–4.1) \times 10^{-3}$	1	0.5	45
	n-C_4H_9Li	$(2.6–6.5) \times 10^{-3}$	1	0.17	46
	s-C_4H_9Li	$(2.6–6.5) \times 10^{-3}$	1	0.17	46
		$(1.5–6.0) \times 10^{-2}$	1	0.5	37
		$(0.7–15) \times 10^{-3}$	1	0.33	37
		$(0.06–16) \times 10^{-3}$	1	0.25	47
	t-C_4H_9Li	$(2.6–6.5) \times 10^{-3}$	1	0.17	46
Benzene	C_2H_5Li	$(5–350) \times 10^{-4}$	1	0.25	41
	n-C_4H_9Li	$(1–10) \times 10^{-3}$	1–2	0	48
		$(3–35) \times 10^{-3}$	1	0.21–0.4	49
	s-C_4H_9Li	$(0.7–10) \times 10^{-3}$	1	0.5	50
	1,1-Diphenylhexyllithium	$(0.1–5) \times 10^{-3}$	1	0.25	51
Toluene	s-C_4H_9Li	$(1.5–4.0) \times 10^{-2}$	1	0.50	37
		$(0.7–15) \times 10^{-3}$	1	0.33	37
		Butadiene			
n-Hexane	n-C_4H_9Li	$(0.9–12) \times 10^{-3}$	1	0.5	1–3
	C_2H_5Li	$(5–100) \times 10^{-4}$	1	0.25	41
n-Heptane	n-C_4H_9Li	$(0.07–6) \times 10^{-3}$	1	0.25	44
		$(0.4–10) \times 10^{-3}$	1	0.167	38
Cyclohexane	s-C_4H_9Li	$(1.5–6.0) \times 10^{-2}$	1	0.50	37
		$(0.7–15) \times 10^{-3}$	1	0.33	37
Benzene	C_2H_5Li	$(5–350) \times 10^{-4}$	1	0.167	41

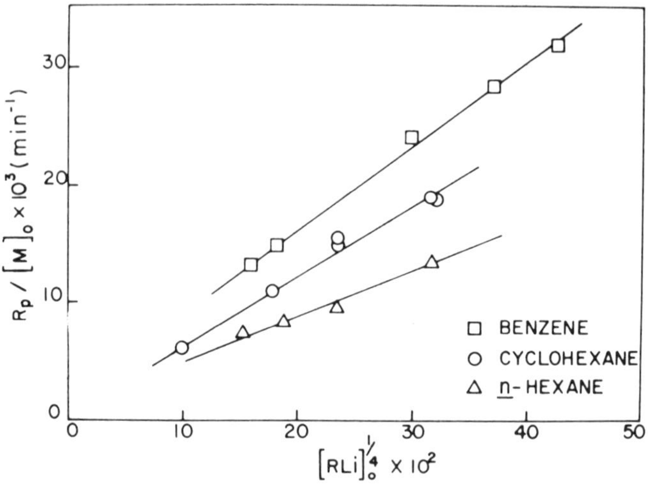

Fig. 4.4 Dependence of propagation rate of isoprene on organolithium chain-end concentration in hydrocarbon solvents at 30°C.[41,41a] $[M]_0 = 2.0$. (Courtesy of Rubber Division, American Chemical Society.)

propagation in the polymerization of styrene, isoprene, and butadiene by organolithium is shown in Table 4.2. The lack of agreement between different investigators, even for the same systems, is quite obvious. Our own experience in this matter is quite illuminating, and may help to explain some of the anomalies. Thus, in our earlier investigations,[1-3] a one-half-order dependence on initiator concentration was found for all those monomers, i.e., styrene, isoprene, and butadiene, in hydrocarbon media. However, in later studies,[41] where greater precautions were taken to eliminate impurities, it was found that, although styrene still showed a one-half-order dependency, this was not the case for the two dienes. Thus, isoprene exhibited approximately one-fourth-order kinetics in three different solvents (see Fig. 4.4), while butadiene obeyed both one-fourth- and even one-sixth-order kinetics, depending on the solvent used (Figs. 4.5 and 4.6). Apparently the earlier findings of one-half-order kinetics were due to a misleading slowdown in rates caused by loss of initiator at the *very low* concentrations, where traces of impurities can have a noticeable effect.

It appears, therefore, that there is reasonable agreement between several different laboratories that in nonpolar media the styrene propagation reaction in organolithium systems obeys one-half-order kinetics while butadiene and isoprene both show a much lower kinetic order. As mentioned previously, by analogy with the kinetics in polar solvents, this was suggestive of an association–dissociation phenomenon in which the dissociated species was the *sole* reactive entity. It is not surprising, therefore, that it was proposed that the

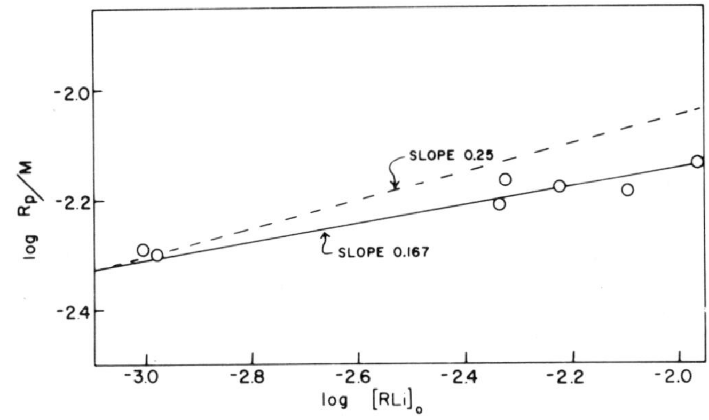

Fig. 4.5 Dependence of propagation rate of butadiene on organolithium chain-end concentration in benzene at 30°C. $[M]_0 = 2$.

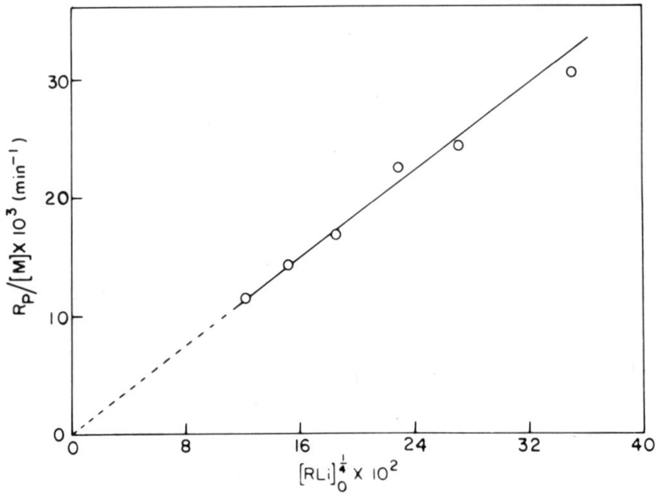

Fig. 4.6 Dependence of propagation rate of butadiene on organolithium chain-end concentration in *n*-hexane at 30°C. $[M]_0 = 2.0$.

kinetic order actually reflected the state of *association* of the active polymer chains,[47] i.e., dimeric for styrene, tetrameric for isoprene and butadiene, and even hexameric for the latter in some solvents. However, as will be discussed in Chapter 5, actual physical measurements of this association of the active chains clearly disprove such an assumption, based as it is on purely kinetic grounds, and show that these polymer chains are always *dimeric*, at least in the polymerization range of concentration.

The inevitable conclusion, then, is that, at least in the case of the organolithium polymerization of the 1,3-dienes, the situation with regard to the propagation step is analogous to that of the initiation reaction. In other words, the monomer most likely first reacts with the associated active chain ends by a coordination–rearrangement mechanism, and the kinetic orders observed are an artifact of the effect of concentration on the character and reactivity of the associated active centers.

2. Polar Vinyl Monomers

Vinyl monomers with heteroatom-containing substituents, such as the acrylates, acrylonitrile, and vinyl chloride, fall into the class of "polar" monomers. As pointed out in Chapter 3, strong bases such as organoalkalies undergo various side reactions with the polar substituents of the monomers, complicating the initiation reaction. This can also apply to the propagation reaction, where such side reactions represent a termination process. It is not surprising, therefore, that little information is available about the kinetics of propagation of these monomers. Certainly none of these systems is understood as well as those involving styrene and the dienes. The reader is therefore referred to Chapter 2, where all recorded work on the various vinyl monomers is tabulated for convenience.

3. Heterocyclics

The propagation reaction in the ring-opening polymerization of heterocyclic monomers differs considerably from that of the vinyl monomers. The anionic mechanism is operative only because of the electronegative character of the heteroatom (e.g., oxygen or sulfur), so that the active chain end is headed by that same heteroatom. Thus, for example, the growing chain in the polymerization of a cyclic oxide is an alkoxide, which is a considerably weaker base than a carbanion. It is not surprising, therefore, that such propagation reactions are considerably slower than those of the vinyl monomers, and generally require the more electronegative counterions (e.g., K^+, Cs^+) for reasonable rates.

The propagation rates in the polymerization of ethylene oxide by the alkali naphthalene complexes were determined by Morton and Dudek,[52] who showed that these were indeed termination-free systems, as shown by the stoichiometry of the molecular weights. They found the propagation reaction to be first order with respect to monomer and fractional order with respect to growing chain ends, the latter order being one-half in the case of the sodium counterion and between one-half and unity for potassium. This was taken to indicate that, in the case of the sodium, only the free anion is active, while for the potassium, both the anion and ion pair can propagate, as in Eq. (4.1).

Similar results were obtained with potassium *t*-butoxide as initiator.[53] An analysis of the kinetics of these reactions is complicated by the fact that the growing poly(ethylene oxide) chains apparently are associated to some extent even in polar solvents such as H_4-furan, just as in the case of vinyl monomers in hydrocarbon media. Yet the active chain-end ion pairs can also undergo an ionic dissociation, the ionization constant reportedly reaching values as high as 10^{-2} in highly polar solvents such as dimethyl sulfoxide.[54] Depending on the counterion and solvent, the kinetic order relative to active centers may be fractional or first, and on this basis, propagation rate constants have been calculated[53-56] for ion pairs or free ions.

As described previously, some of the cyclic sulfides are also amenable to anionic polymerization. Here again, association of chain ends as well as ionic dissociation, has been observed[57-59] in the polymerization of propylene sulfide in polar solvents such as H_4-furan. Determination of the ionization constants in these cases has made it possible to calculate propagation rate constants[58,59] for both the free anions and some of the ion pairs. The special case of organolithium initiation has already been discussed; here both ethylene and propylene sulfides propagate as lithium thiolate chain ends, while trimethylene sulfide (thiacyclobutane) propagates as a lithium carbanion chain end.[60,61]

An entirely different but no less important heterocyclic monomer system is that of the siloxanes. This work has been largely limited to the popular and commercially used cyclic tetramer, octamethylcyclotetrasiloxane, usually referred to as D_4. This type of polymerization, as mentioned previously, is initiated by strong bases, e.g., KOH, and has the character of an equilibrium polymerization, due to the very low enthalpy of this ring-opening reaction.[62,63] Thus, at about 150°C in the undiluted monomer, the equilibrium concentration[64] of D_4 is about 6%. The kinetics of this base-catalyzed polymerization were shown[65] to be those of a termination-free system with the propagating species being the free silanolate ion as indicated by the one-half-order dependence of rate on initiator concentration. This was confirmed by a study[64] of the effect of polar solvents, which increased the rate of polymerization by presumably increasing the ionic dissociation of the potassium silanolate chain ends.

Although the tetramer D_4 requires a base at least as strong as KOH for reasonable polymerization rates, it was later found that the cyclic trimer hexamethylcyclotrisiloxane (D_3), was susceptible to polymerization by organolithium initiators,[66] presumably because of the much greater ring strain in this monomer. As mentioned previously, this leads to an interesting phenomenon, i.e., the synthesis of polysiloxanes having a very narrow MWD, which is not possible to achieve with the D_4 tetramer, since the potassium silanolate chain ends participate in bond interchange with the linear polysiloxane. The lithium silanolate is apparently too weak a base to cause any substantial amount of such bond interchange, at least within the time limit of the polymerization reaction. Such organolithium systems can also be used for the synthesis of precisely controlled block copolymers with styrene or other vinyl monomers.[67–69]

References

1. M. Morton, E. E. Bostick, and R. A. Livigni, *Rubber Plast. Age* **42**, 397 (1961).
2. M. Morton, E. E. Bostick, R. A. Livigni, and L. J. Fetters, *J. Polym. Sci., Part A* **1**, 1735 (1963).
3. M. Morton, L. J. Fetters, and E. E. Bostick, *J. Polym. Sci., Part C* **1**, 311 (1963).
4. Y. L. Spirin, D. K. Polyakov, A. R. Gantmakher, and S. S. Medvedev, *Dokl. Akad. Nauk SSSR* **139**, 899 (1961); *CA* **56**, 2551b (1961).
5. S. Bywater and D. J. Worsfold, *Can. J. Chem.* **45**, 1821 (1967).
6. G. Allen, G. Gee, and C. Stretch, *J. Polym. Sci.* **48**, 189 (1960).
7. C. Stretch and G. Allen, *Polymer* **2**, 151 (1961).
8. C. Geacintov, J. Smid, and M. Szwarc, *J. Am. Chem. Soc.* **83**, 1253 (1961); **84**, 2508 (1962); M. Szwarc, "Carbanions, Living Polymers and Electron Transfer Processes," p. 408. Wiley (Interscience), New York, 1968.
9. D. J. Worsfold and S. Bywater, *J. Chem. Soc.* p. 5234 (1960).
10. H. Hostalka, R. V. Figini, and G. V. Schulz, *Makromol. Chem.* **71**, 198 (1964).
11. H. Hostalka and G. V. Schulz, *Z. Phys. Chem. Wiesbaden* [N. S.] **45**, 286 (1965).
12. D. N. Battacharyya, C. I. Lee, J. Smid, and M. Szwarc, *Polymer* **5**, 54 (1964); *J. Phys. Chem.* **69**, 612 (1965).
13. T. Shimomura, K. J. Tölle, J. Smid, and M. Szwarc, *J. Am. Chem. Soc.* **89**, 796 (1967).
14. T. Shimomura, J. Smid, and M. Szwarc, *J. Am. Chem. Soc.* **89**, 5743 (1967).
15. A. Parry, J. E. L. Roovers, and S. Bywater, *Macromolecules* **3**, 355 (1970).
16. G. Löhr and S. Bywater, *Can. J. Chem.* **48**, 2031 (1970).
17. D. N. Battacharyya, J. Smid, and M. Szwarc, *J. Phys. Chem.* **69**, 624 (1965).
18. M. Szwarc, "Carbanions, Living Polymers and Electron Transfer Processes," p. 421. Wiley (Interscience), New York, 1968.
19. F. S. Dainton, G. C. East, G. A. Harpell, N. R. Hurwurth, K. J. Ivin, R. T. LaFlair, R. H. Pallen, and K. M. Hui, *Makromol. Chem.* **89**, 257 (1965).
20. L. Böhm, W. K. R. Barnikol, and G. V. Schulz, *Makromol. Chem.* **110**, 222 (1967).
21. B. L. Funt and S. N. Bhadani, *J. Polym. Sci., Polym. Chem. Ed.* **9**, 2429 (1968).
22. F. S. Dainton, K. M. Hui, and K. J. Ivin, *Eur. Polym. J.* **5**, 387 (1969).
23. F. S. Dainton, G. A. Harpell, and K. J. Ivin, *Eur. Polym. J.* **5**, 395 (1969).
24. K. M. Hui and T. L. Ng, *J. Polym. Sci., Polym. Chem. Ed.* **7**, 3101 (1969).
25. J. Comyn and K. J. Ivin, *Eur. Polym. J.* **5**, 587 (1969).
26. J. Comyn, F. S. Dainton, and K. J. Ivin, *Eur. Polym. J.* **6**, 319 (1970).
27. M. Nakayama, H. Hirohara, K. Takaya, and N. Ise, *J. Polym. Sci., Polym. Chem. Ed.* **8**, 3653 (1970).
28. H. Hirohara, M. Nakayama, R. Kawabata, and N. Ise, *J. Chem. Soc., Faraday Trans. 1* **68**, 51, 58 (1972).

29. K. Takaya, H. Hirohara, M. Nakayama, and N. Ise, *Trans. Faraday Soc.* **67**, 119 (1971).
30. M. Fisher and M. Szwarc, *Macromolecules* **3**, 23 (1970).
31. M. Tardi and P. Sigwalt, *Eur. Polym. J.* **8**, 137, 151 (1972).
32. J. E. L. Roovers and S. Bywater, *Trans. Faraday Soc.* **62**, 701 (1966).
33. D. J. Worsfold and S. Bywater, *Can. J. Chem.* **38**, 1891 (1960).
34. K. F. O'Driscoll and A. V. Tobolsky, *J. Polym. Sci.* **35**, 259 (1959).
35. F. J. Welch, *J. Am. Chem. Soc.* **81**, 1345 (1959); **82**, 6000 (1960).
36. R. C. P. Cubbon and D. Margerison, *Proc. R. Soc., London, Ser. A* **268**, 260 (1962); *Polymer* **6**, 102 (1965).
37. H. Hsieh, *J. Polym. Sci., Part A* **3**, 153, 173 (1965).
38. A. F. Johnson and D. J. Worsfold, *J. Polym. Sci., Part A* **3**, 449 (1965).
39. L. J. Fetters, *J. Res. Natl. Bur. Stand., Sect. A* **69**, 159 (1965).
40. A. Korotkov, N. N. Chesnokova, and L. B. Truchmanova, *Vysokomol. Soedin.* **1**, 46 (1959); *Polym. Sci. USSR (Engl. Transl.)* **1**, 10 (1960); *Rubber Chem. Technol.* **33**, 610 (1960).
41. M. Morton, R. A. Pett, and J. F. Fellers, *Prepr. IUPAC Macromol. Symp.*, Vol. 1, p. 69 (1966).
41a. M. Morton and L. J. Fetters, *Rubber Chem. Technol.* **48**, 359 (1975).
42. I. Kuntz, *J. Polym. Sci.* **A2**, 2827 (1964).
43. C. Lundborg and H. Sinn, *Makromol. Chem.* **41**, 242 (1960); H. Sinn and C. Lundborg, *ibid.* **47**, 86 (1961); H. Sinn and O. T. Onsager, *ibid.* **52**, 246 (1962); H. Sinn and W. Hoffman, *ibid.* **56**, 234 (1962); H. Sinn and F. Banderman, *ibid.* **62**, 134 (1963); H. Sinn, C. Lundborg, and O. T. Onsager, *ibid.* **70**, 222 (1964); W. Gerber, J. Hinz, and H. Sinn, *ibid.* **144**, 97 (1971).
44. Yu. L. Spirin, A. R. Gantmakher, and S. S. Medvedev, *Dokl. Akad. Nauk SSSR* **146**, 368 (1962); *CA* **58**, 1537d (1963); Yu. L. Spirin, D. K. Polyakov, A. R. Gantmakher, and S. S. Medvedev, *J. Polym. Sci.* **53**, 233 (1961).
45. B. Francois, V. Sinn, and J. Parrod, *J. Polym. Sci., Part C* **4**, 375 (1964).
46. A. Guyot and J. Vialle, *J. Macromol. Sci., Chem.* **A4**, 107 (1970).
47. D. J. Worsfold and S. Bywater, *Can. J. Chem.* **42**, 2884 (1964).
48. J. Minoux, B. François, and C. Sadron, *Makromol. Chem.* **44–46**, 519 (1961); J. Minoux, *Rev. Gen. Caoutch.* **39**, 779 (1962); *Makromol. Chem.* **61**, 22 (1963).
49. D. Margerison, D. M. Bishop, G. C. East, and P. McBride, *Trans. Faraday Soc.* **64**, 1872 (1968).
50. D. N. Cramond, P. S. Lawry, and J. R. Urwin, *Eur. Polym. J.* **2**, 107 (1966).
51. J. M. Alvarino, A. Bello, and G. M. Guzman, *Eur. Polym. J.* **8**, 53 (1972).
52. T. Dudek, Ph.D. Dissertation, University of Akron, Akron, Ohio, 1961.
53. C. C. Price and D. D. Carmelite, *J. Am. Chem. Soc.* **88**, 4039 (1966).
54. A. A. Solovyanov and K. S. Kazanskii, *Vysokomol. Soedin., Ser. A* **12**, 2114 (1970); **14**, 1063, 1072 (1972).
55. C. E. H. Bawn, A. Ledwith, and N. McFarlane, *Polymer* **10**, 653 (1969).
56. K. S. Kazanskii, A. A. Solovyanov, and S. G. Entelis, *Eur. Polym. J.* **7**, 1421 (1971).
57. P. Guerin, S. Boileau, and P. Sigwalt, *Eur. Polym. J.* **7**, 1119 (1971).
58. P. Hemery, S. Boileau, and P. Sigwalt, *Eur. Polym. J.* **7**, 1581 (1971).
59. G. Tersac, S. Boileau, and P. Sigwalt, *J. Chim. Phys.* **65**, 1141 (1968); *Makromol. Chem.* **149**, 153 (1971).
60. M. Morton and R. F. Kammereck, *J. Am. Chem. Soc.* **92**, 3217 (1970).
61. M. Morton, R. F. Kammereck, and L. J. Fetters, *Macromolecules* **4**, 11 (1971); *Br. Polym. J.* **3**, 120 (1971).
62. D. W. Scott, *J. Am. Chem. Soc.* **68**, 2294 (1946).
63. J. B. Charmichael and R. Winger, *J. Polym. Sci., Part A* **3**, 971 (1965).
64. M. Morton and E. E. Bostick, *J. Polym. Sci., Part A* **2**, 523 (1964).
65. W. T. Grubb and R. C. Osthoff, *J. Am. Chem. Soc.* **77**, 1405 (1955).

66. J. G. Saam, D. J. Gordon, and S. Lindsey, *Macromolecules* **3**, 1 (1970).
67. W. G. Davies and D. P. Jones, *Polym. Prepr., Am. Chem. Soc., Div. Polym. Chem.* **11**, 447 (1970); "Colloidal and Morphological Behavior of Block and Graft Copolymers." Plenum, New York, 1971.
68. A. Ward, T. Kendrick, and J. G. Saam, *Polym. Prepr., Am. Chem. Soc., Div. Polym. Chem.* **15** (1), 183 (1974); *in* "Copolymers, Polyblends, and Composites" (N. A. S. Platzer, ed.) p. 300. American Chemical Society, Washington, D.C., 1975.
69. M. Morton, Y. Kesten, and L. J. Fetters, *Polym. Prepr., Am. Chem. Soc., Div. Polym. Chem.* **15** (2), 175 (1974); *Appl. Polym. Symp.* **26**, 113 (1975).

Chapter

5 Association Phenomena in Organolithium Polymerization

1. Association of Propagating Chain Ends in Nonpolar Solvents

In the case of *polar solvents*, evidence concerning the nature of the propagating anionic chain was obtained from reaction kinetics as well as by direct physical measurements. Hence there appears little doubt that the growing chain end can be either a free anion or some type of ion pair. However, studies of the anionic polymerization of such monomers as styrene, butadiene, and isoprene in *nonpolar solvents*, which were largely restricted to organolithium systems, led to much more confusion about the nature of the propagating chain end. Thus the fractional kinetic orders obtained for these monomers led to speculations and conjectures[1] relating the kinetic order to the association of growing chain ends.

a. Determination of Association by Viscometry

Actual physical measurements of association of these growing polymer chains were first carried out[2,3] by utilizing the relation between the viscosity of polymer solutions and their molecular weights. As can be seen, the data shown in Table 3.2 on the state of association of organolithium initiators were obtained by the usual cryoscopic or vapor pressure methods for determination of molecular weight. These would also require high concentrations of organolithium, several orders of magnitude greater than those used for synthesis of high polymers. It is obvious, therefore, that such methods cannot be applied with any degree of accuracy to the determination of the molecular weight of the polymers. On the other hand, any association of macromolecules leads to dramatic differences in molecular weight, i.e., the doubling, tripling, etc., of an already high molecular weight species. Hence any molecular weight method suitable for macromolecules should be applicable. The problem is, of course, considerably simplified in the case of these anionic nonterminating polymerizations, since it is possible, at the conclusion of the polymerization, to "terminate" the active chain ends (by addition of water or alcohol) and thus destroy the association. In other words, the state of association is measurable by determining the molecular weight of the chains before and after termination.

Since laboratory polymerizations of this type must be carried out in rigorously pure, sealed systems, the molecular weight method is best applied *in situ*, with a minimum of transfer operations. Furthermore, since these polymerizations generally result in 10% or 20% solutions of the polymer, the most attractive molecular weight method would appear to be based on viscosity measurements, which are particularly sensitive in this range of concentration. The association can then be quantitatively determined by the simple expedient of measuring the viscosity of the sealed solution before and after termination. Since the latter is accomplished by adding one or two drops of alcohol to several hundred milliliters of solution, the concentration of polymer can be considered as constant.

The viscosity of polymer melts or "concentrated" solutions (as opposed to "dilute" solutions, i.e., $< 1\%$) has been studied both theoretically[4] and experimentally.[5,6] From theoretical considerations it has been predicted that the viscosity of *linear* macromolecules should be proportional to the 3.5 power of the molecular weight, but the best experimental data[5,6] actually indicated a slightly lower exponent:

$$\eta = KM_w^{3.4} \tag{5.1}$$

where η is the viscosity of solution, K a constant which includes the concentration of the solution, and M_w the weight-average molecular weight. As indicated, Eq. (5.1) applies both to melts and "concentrated" solutions, but, for the latter, this equation can only be used provided the following relation[6] holds:

$$v_2 M_w \geqslant 2M_e \tag{5.2}$$

where v_2 is the volume fraction of polymer in solution and M_e the molecular weight between entanglements of the polymer chains.

In applying Eq. (5.1) to the measurement of association, a vacuum capillary viscometer[2,3] was used to measure the flow times of the "active" and "terminated" polymer solutions. Assuming *two* linear chains associated at their *ends*, the following relation should hold:

$$t_a/t_t = N^{3.4} \tag{5.3}$$

where t_a is the flow time of "active" polymer solution, t_t the flow time of "terminated" polymer solution, and $N = M_w(\text{active})/M_w(\text{terminated})$ the association number. The parameter N should therefore be very close to 2, since it represents the number of chains (on a weight-average basis)* which are associated together. N should actually be slightly less than 2 if allowance is made for a dissociation equilibrium. Since $2^{3.4} = 10.56$, the ratio of t_a/t_t is greater than 10, representing a tremendous decrease in viscosity of the polymer solution after termination. Typical values for these measurements[2] are listed in Table 5.1 for polystyrene, polyisoprene, and polybutadiene, showing how sensitive these viscosity measurements are to such association effects. Flow times are also shown for two polymer solutions in a polar solvent (H_4-furan) in order to demonstrate the absence of any association.

TABLE 5.1 Viscosity and Association of Organolithium Polymers[a]

Polymer	Solvent	Flow time (sec): Active	Flow time (sec): Terminated	N
Polystyrene	Benzene	5770	560	1.99
	Benzene	825	81.1	1.98
Polyisoprene	*n*-Hexane	836	81.1	1.99
	n-Hexane	1060	103.5	1.98
	H_4-furan	3150	3130	1.00
Polybutadiene	Hexane	1030	75.5	2.16
	H_4-furan	2580	2570	1.00

[a] At 25°C. From Morton *et al.*[2] and Morton and Fetters[3].

The case of polybutadiene deserves special attention, since it shows an N value *above* 2. Other measurements[2,3] also corroborated that this polymer showed N values in hexane ranging between 2 and 3, indicating states of association higher than 2. This apparently anomalous result can, however, be

* N should, of course, actually be obtained from the ratio of *number-average* molecular weights. However, since these carefully prepared polymers can be made to approach a monodispersity of molecular weights (Poisson distribution), a ratio of $M_w/M_n = 1.05$ or less is quite common, and the error involved in N values is quite small.

easily explained by the finding[2,3] that some slight degree of crosslinking of polymer chains occurs during the latter stages of butadiene polymerization, but not in the case of isoprene or styrene. Such a crosslinking reaction during the polymerization will, of course, lead to polyfunctional chains, e.g., dilithium or higher, which could then associate with *more than one* other chain, thus leading to values of N greater than 2.

It can be seen from Table 4.2 that the propagation reaction had been studied in several different nonpolar solvents, whereas the data in Table 5.1 are limited to a single solvent for each species. In order to coordinate these results, later studies[7,8] were carried out on the association of these propagating chains in the several nonpolar solvents used by various investigators for kinetic measurements. Some of these viscometric data are reproduced in Table 5.2, which also shows the concentrations used of initiator (RLi) and monomer (M). It is clear that both the polystyryl- and polyisoprenyllithium species are associated in *pairs* in all the solvents used.

TABLE 5.2 Viscosity and Association of Polystyryllithium and Polyisoprenyllithium[a]

Polymer	Solvent	[RLi] ($\times 10^3$)	[M]	Flow time (sec) Active	Flow time (sec) Terminated	N
Polystyrene	Benzene	1.4	1.8	1960	195.8	1.97
Polyisoprene	Benzene	1.8	3.4	1268	126.0	1.97
	Cyclohexane	0.9	2.9	169	16.6	1.98
	n-Hexane	1.4	3.4	690	67.5	1.98
	n-Hexane/benzene 60/40 (vol.)	1.2	3.2	1190	114	1.99
	Cyclohexane	2.3	4.1	7000	664	2.0
	Cyclohexane	2.3	3.0	2380	227	2.0

[a] At 30°C. From Morton *et al.*[7] and Fetters and Morton.[8]

Although the exponent of 3.4 in Eq. (5.1) had been carefully evaluated for various polymers,[5,6] subject to the conditions set by Eq. (5.2), some additional verification was obtained on the actual polymers (polystyrene and polyisoprene) used in these association studies, after termination. For this purpose, the viscosity-average molecular weights M_v were calculated from intrinsic viscosity data, since these would virtually represent the weight-average molecular weights for these near-monodisperse polymers. It should be noted that the intrinsic viscosity equations used in these cases had previously been derived specifically* for these anionic polymers. The results are shown in Table 5.3 and corroborate the validity of both the experimental techniques and the

* For polystyrene, see Altares *et al.*[9] For polyisoprene, see Morton *et al.*[10]

TABLE 5.3 Verification of Exponent *a* in Equation (5.1)[a]

Polymer	Solvent	$M_v(\times 10^{-4})$	v_2	Flow time	*a*
Polystyrene	Benzene	9.8	40	58.0 min	3.41
	Benzene	12.8	40	147.0 min	
	Benzene	7.7	50	51.5 min	3.39
	Benzene	12.8	50	284.0 min	
Polyisoprene	*n*-Hexane	29.3	3.7	66.5 sec	3.38
	n-Hexane	109.7	3.7	5700 sec	
	n-Hexane	132.0	3.7	10,740 sec	3.42

[a] From Morton and Fetters[3] and Morton *et al.*[7]

use of Eq. (5.1). In recent years work in other laboratories[11–14] on this type of near-monodisperse polystyrene and polyisoprene has produced a wealth of additional experimental evidence which confirms the validity of Eq. (5.1).

b. The "Capping" Method

As mentioned previously, in connection with the data in Table 5.1, it was not possible to determine the association of polybutadienyllithium chains because of some extent of crosslinking which seemed to occur during the polymerization. To provide direct evidence for the dimeric state of association of polybutadienyllithium, a special "capping" technique was designed.[7] This involved the polymerization of styrene to completion, followed by the addition of a very small amount of butadiene, sufficient to add a few units of the latter monomer to the end of the polystyrene chain, but not enough to introduce any serious possibility of crosslinking. Viscosity measurements were then carried out on the "capped" polymers, as usual, before and after termination. Since it is well known from copolymerization[15] and other[16] studies that the "cross-over" reaction of polystyryllithium and butadiene (or isoprene) is very rapid, one can be reasonably certain of the complete conversion of the polystyryllithium to polydienyllithium chain ends, provided there is a sufficiently high molar ratio of diene monomer to active chain ends.

Some results obtained with this capping technique,[7] using both butadiene and isoprene, are shown in Table 5.4. It is clear that the polybutadienyllithium chains ends, uncomplicated by crosslinking reactions, are associated in *pairs*, just as in the case of polystyryl- and polyisoprenyllithium. A similar conclusion was reached by Makowski and Lynn[17] who also used a viscometric technique to measure the state of association of *low molecular weight* polybutadienyllithium. Table 5.4 also shows that the isoprene-capped chains, too, are in a

TABLE 5.4 Association of "Capped" Polystyryl Lithium[a]

Solvent	[RLi] ($\times 10^3$)	[M]	Diene[b]	Flow time (sec) Polystyryl lithium	"Capped" polymer Active	Terminated	*N*
Benzene	0.60	1.5	Butadiene	—	398.8	38.4	1.99
Benzene	0.63	1.5	Butadiene	—	268.9	26.2	1.98
Benzene	2.40	3.2	Isoprene	1899	1974	187.5	2.00
Cyclohexane	2.40	3.0	Isoprene	1567	1591	158.4	1.97

[a] At 30°C. From Morton *et al.*[7]
[b] Approximately 2000 g of diene was added per mole of chain ends.

dimeric association state, just as in the case of the polyisoprenyllithium. One additional result of these experiments is that they clearly demonstrate that the polyisoprenyllithium *cannot* be associated in any higher form than the dimer, since there is no significant change in viscosity after capping of the polystyryllithium by the isoprene. If, as has been claimed purely on kinetic grounds, polyisoprenyllithium is associated as a tetramer[1,18] (and polybutadienyllithium as a hexamer[15,19]),* then the conversion of a dimeric polystyryllithium to a tetrameric polyisoprenyllithium should have been accompanied by a noticeable increase in viscosity, which obviously did not occur.

In a very recent publication, Al-Jarrah and Young[21] extended this same "capping" technique to a study of the association of a series of substituted polydienyllithiums in *n*-hexane at 25°C. They found that those monomers which could be assumed to lead to chain ends of a primary carbanionic structure were consistently associated as dimers. These included 2,3-dimethylbutadiene, 1,3-pentadiene, 2-methyl-1,3-pentadiene, 3-methyl-1,3-pentadiene, 4-methyl-1,3-pentadiene, and 2,4-dimethyl-1,3-pentadiene. However, in the case of 2,4-hexadiene and 1,3-cyclohexadiene, which have been shown[22] to lead to secondary carbanionic chain ends, a substantial dissociation of the dimeric chain ends was found in corroboration of previous findings.[22] The latter phenomenon will be discussed in a later chapter.

The type of viscosity increase that would have occurred in Table 5.4, had there been an increase in association state after capping, was amply demonstrated[8] by using chemical linking agents to bond together two, three, or four chain ends. Various silicon halides were used for this purpose because of their known reactivity with organolithium. The results are shown in Table 5.5

* An attempt was actually made[20] to obtain molecular weight data supporting the tetrameric state of association of polyisoprenyllithium, but these were shown to be based on faulty measurements.[8]

TABLE 5.5 Viscosities of Associated and Linked Polydienes in Cyclohexane[a]

Polymer	Linking agent	Flow time (sec) Active	Flow time (sec) Linked	Flow time (sec) Terminated	Mol wt[b] ($M_n \times 10^{-5}$) Primary	Mol wt[b] ($M_n \times 10^{-5}$) Linked[c]
Polyisoprene	$(CH_3)_2SiCl_2$	235	229	229	0.85 ± 0.4	1.6 ± 0.1
	CH_3SiCl_3	500	825	820	1.35 ± 0.1	3.6 ± 0.3
Butadiene-capped polyisoprene	$SiCl_4$	2480	4800	4770	1.6 ± 0.1	6.0 ± 0.4

[a] 25°C, $v_2 = 0.18$–0.20. From Fetters and Morton.[8]
[b] By osmometry.
[c] Gel permeation chromatography indicated that these polymers contained 2–5% of unlinked chains, presumably due to experimental difficulties with stoichiometry.

for polyisoprenyllithium as well as for polybutadienyllithium. It is clearly seen that the *difunctional* linking agent, dimethyldichlorosilane, caused no noticeable change in viscosity, while both the *tri-* and *tetrafunctional* silanes caused a very substantial increase, as might be expected when three and four chain ends, respectively, are linked together to form a "star-branched" polymer. The fact that there was a negligible decrease in flow time after termination in all cases shows that there was little or no unlinked polymer. This was, of course, also corroborated by the M_n values as well as by the gel permeation chromatography.

c. Determination of Association by Light Scattering

Although the viscometric studies described heretofore offer very convincing evidence for the dimeric association of these organolithium polymers in nonpolar solvents, this is by no means the only such evidence available. Light scattering measurements[7] have also been used as a means of determining the molecular weight of polyisoprenyllithium. As in the case of the viscometric

TABLE 5.6 Light Scattering of Polyisoprenyl Lithium[a]

Polymer no.	Zimm plot intercept $C/(\Delta I \times \Phi)$ Active	Zimm plot intercept $C/(\Delta I \times \Phi)$ Terminated	Mol wt ($M_w \times 10^{-5}$) Active	Mol wt ($M_w \times 10^{-5}$) Terminated
1	4.6	9.6	2.8	1.4
2	5.0	9.5	2.6	1.4
2	5.1	9.6	2.6	1.4

[a] In *n*-hexane at 27°C. From Morton *et al.*[7]

studies, a suitable light scattering cell had to be built into the high vacuum apparatus in order to determine the scattering both before and after termination of the chain ends, and these experimental procedures have been described.[7] A summary of the light scattering data and molecular weights is shown in Table 5.6. It can be seen at once that the process of "termination" of the polyisoprenyllithium chains, i.e., the hydrolysis of the carbon–lithium bond by addition of methanol, leads to a reduction by half of the measured molecular weight, hence the inescapable conclusion is that the polyisoprenyllithium chain ends are associated *in pairs*, which confirms and reinforces the results of the viscosity measurements described previously.

It should be mentioned in this connection that some claims have been made that light scattering studies showed evidence for the tetrameric (and trimeric) association of polyisoprenyllithium[1,20] and the tetrameric association of polybutadienyllithium.[20] However, these conclusions were later shown[7,8] to be invalid, since, in one case[1] there appeared to be inconclusive evidence for association numbers ranging from 1.6 to 3.7, while in the other case,[20] the choice of an unsatisfactory solvent led to dubious results.

In conclusion, therefore, all the reliable experimental evidence obtained to date clearly indicates the dimeric association of polystyryl-, polyisoprenyl-, and polybutadienyllithium in hydrocarbon media at concentrations in the *macromolecular range*, i.e., $< 10^{-2}$ M. This conclusion is supported by the data reported by all the laboratories with one exception,[20] including not only our own work[2,3,7] but that of three other groups,[17,21,23,24] using viscosity,[17,21] cryoscopy,[23] and laser light scattering[24] measurements.

2. Effect of a Polar Solvent on Association of Organolithium Polymers in Nonpolar Solvents

As indicated in Table 5.1, the active chain ends in the organolithium polymerization of styrene, isoprene, and butadiene show no signs of association in a polar solvent such as H_4-furan. This is undoubtedly due to the ability of such ethers to solvate the lithium cation, leading to charge separation, delocalization, and stabilization of the ionic bond, as indicated in a later chapter. Because of these effects, there appears to be no need to stabilize the carbon–lithium bond by molecular association, and none is observed. In addition, of course, as indicated earlier, the higher dielectric constant of the polar solvents permits a dissociation of the ionic bond (general solvent effect).

In this connection, it is interesting to study the "specific" solvent effect which results from the solvation of the lithium cation by the polar solvent, and this can be done by observing the *change* in the state of association of the polymer–lithium chain ends caused by the addition of stoichiometric (and

TABLE 5.7 Effect of H_4-Furan on Polystyryllithium Association in Benzene[a]

[RLi] ($\times 10^3$)	$[C_4H_8O]/[RLi]$	Flow time (sec) Active	Flow time (sec) Terminated	N
1.4	0	1960	195.8	1.97
1.2	2.6	347.5	86.0	1.51
1.2	3.1	514.4	129.3	1.50
0.62	4.3	200.0	62.6	1.41
3.8	7.4	242.0	179.6	1.09

[a] At 30°C.

TABLE 5.8 Effect of H_4-Furan on Polyisoprenyllithium Association in n-Hexane[a]

[RLi] ($\times 10^3$)	$[C_4H_8O]/[RLi]$	Flow time (sec) Active	Flow time (sec) Terminated	N
1.40	4.7	441	50.7	1.87
1.40	12.2	723	121.5	1.69
3.20	20.4	3980	1025	1.49
1.40	50.5	190.5	65.3	1.37
1.30	350	163	125	1.08
1.50	560	784	665	1.04

[a] At 30°C.

higher) amounts of the polar solvent. Such measurements were carried out using the same viscometric method as described previously for the case of polystyrene[7] and polyisoprene.[3] Some of the data are reproduced in Table 5.7 for polystyrene and in Table 5.8 for polyisoprene. It can be seen that the H_4-furan is much more effective in disrupting the association of the polystyryllithium than the polyisoprenyllithium, only trace amounts ($[C_4H_8O]/[RLi] \sim 10$) being sufficient in the former case, while much larger concentrations are needed in the latter case. Furthermore, it is obvious that this solvation reaction is not stoichiometric but appears to participate in an equilibrium.

3. Association and Solvation Equilibria

Since the association phenomenon described above must obviously involve an equilibrium between associated and unassociated species, it should be possible to determine the equilibrium constants, provided the experimental methods are sufficiently accurate. The values of N shown in Tables 5.1, 5.2, and

5.4 are very close to 2, indicating that any equilibrium strongly favors the associated state. Hence the determination of the amount of unassociated species would require very accurate molecular weight measurements. Actually the viscometric method defined by Eq. (5.1) is sufficiently precise to provide such information, thus enabling the calculation of an equilibrium constant. This was actually done in the case of polyisoprene[3] in n-hexane at several temperatures leading to calculated values of K_e (dissociation) of about 10^{-7} and a corresponding heat of dissociation, ΔH, of about 37 kcal. Such values appear to be quite reasonable for association* of organolithium compounds.[25,26a] It is interesting to note that, if such a calculation is applied to polystyryllithium in benzene[7] at 30°C, a K_e value of about 10^{-6} is obtained, indicating a somewhat higher dissociation equilibrium for this system.

The same approach can be used in calculating a solvation equilibrium based on the type of data shown in Tables 5.7 and 5.8. Such values were obtained for the solvation of polyisoprenyl-[3] and polystyryllithium[7] by H_4-furan, indicating the equilibrium

$$(RLi)_2 + 2C_4H_8O \rightleftharpoons 2(RLi \cdot C_4H_8O) \tag{5.4}$$

The values of K_e at 30°C were about 1 for the polyisoprene in n-hexane and about 10^2 for the polystyrene in benzene, indicating a very strong solvation effect, especially in the case of the latter. It is interesting to note that this corroborates the fact that this solvation is an equilibrium process and not a simple stoichiometric reaction. Thus the solvation of the polystyryllithium is sufficiently strong to lead to a complete disruption of the associated species at a solvent/lithium ratio of 10 (Table 5.7), while in the case of the polyisoprenyllithium, there is still some degree of association even at a solvent/lithium molar ratio of 600 (Table 5.8).

If the type of data shown in Tables 5.7 and 5.8 are obtained at different temperatures, it should be possible to determine the heat of solvation in the reaction

$$RLi + C_4H_8O \rightleftharpoons RLi \cdot C_4H_8O \tag{5.5}$$

Such measurements were made for polyisoprenyllithium in n-hexane at 30°C and 40°C, and the calculated value of ΔH (solvation) turned out to be about 18 kcal. This appears to be a reasonable value as indicated by recent direct calorimetric measurements.[27]

* It is indeed striking that a very recent publication[26] predicts a value of 34.2 kcal $mole^{-1}$ for the heat of dissociation of ethyllithium dimers to monomers, based on molecular orbital calculations!

4. Effect of Alkyllithium Initiators on Chain-End Association (Cross Association)

Since it is known that the alkyllithium compounds used as initiators are themselves in an associated state in nonpolar solvents (see Table 3.2), the question arises of the possibility of an interaction between the initiator and the growing chains during the initiation period. Such information is of critical importance in any kinetic studies, since any such interaction of these associated species could markedly affect both the initiation and propagation reactions during this period.

TABLE 5.9 Cross Association of Ethyllithium (EtLi) and Polyisoprenyllithium (ILi) in *n*-Hexane[a, b]

[EtLi]/[ILi]	Flow time (sec) Active	Flow time (sec) Terminated	*N*
3.2	225.2	57.2	1.50
5.4	151.2	57.2	1.32
8.0	119.3	57.2	1.24
11.1	101.7	57.2	1.18

[a] At 40°C. From Morton *et al.*[30]
[b] [ILi] = 1.9×10^{-3} *M*.

The simplest way of determining such a cross-association reaction is to note the effect of added initiator to a completely polymerized monomer solution. Qualitative observations of the effect of *n*-butyllithium on the association of polyisoprenyllithium in *n*-hexane were first noted[2] from viscosity measurements. Later similar effects were reported for *sec*-butyllithium[28] and *tert*-butyllithium.[29] However, the first quantitative treatment of such a cross association was carried out[30] on the ethyllithium–polyisoprene system in *n*-hexane. Typical data which show how the addition of ethyllithium affects the viscosity of polyisoprenyllithium solutions are shown in Table 5.9. From the values of *N*, the "association number," it was then possible to determine the values of the unknown parameters x and y in the equation

$$(\mathrm{ILi})_2 + (2x/y)(\mathrm{EtLi})_y \rightleftharpoons 2[\mathrm{ILi} \cdot (\mathrm{EtLi})_x] \qquad (5.6)$$

Equation (5.6) is based on the known dimeric association of the polymer but without any assumptions as to the state of association of the ethyllithium in this system. The equilibrium constants calculated from the N values in Table 5.9 were consistent only for values of $x = 3$, $y = 6$ and showed a wide variation when x was varied from 1 to 4 and y was varied from 1 to 6. Hence the cross-association reaction can be assumed to be represented by

$$(ILi)_2 + (EtLi)_6 \overset{K_e}{\rightleftharpoons} 2(ILi \cdot 3EtLi) \tag{5.7}$$

with the value of K_e being about 6, i.e., indicating a very strong cross-association tendency. It is interesting to note that the proposed hexameric association of the ethyllithium in this system is in excellent agreement with the experimental values found by Brown *et al.*[31,32] for this compound in hydrocarbon solvents at much higher concentrations.

The significance of these results to the kinetics of the polymerization reactions cannot be overlooked. Thus as soon as the initiation reactions get under way, the cross-associated species ILi · 3EtLi will begin to tie up significant quantities of the initiator, and the initiation rate of this new species can be assumed to be different from that of the hexameric associated ethyllithium. Hence it would appear that this would pose great difficulties in the measurement of the true initiation rate, except possibly at the very outset of the reaction. Such considerations can therefore raise serious doubts about the validity of initiation kinetics reported in the literature.

References

1. D. J. Worsfold and S. Bywater, *Can. J. Chem.* **38**, 1891 (1960); **42**, 2884 (1964).
2. M. Morton, E. E. Bostick, and R. A. Livigni, *Rubber Plast. Age* **42**, 397 (1961).
3. M. Morton and L. J. Fetters, *J. Polym. Sci., Part A* **2**, 3311 (1964).
4. F. Bueche, *J. Chem. Phys.* **25**, 599 (1956).
5. V. R. Allen and T. G. Fox, *J. Chem. Phys.* **41**, 337, 344 (1964).
6. T. G. Fox, B. S. Gratch, and S. Loshaek, *in* "Rheology" (F. Eirich, ed.), p. 443. Academic Press, New York, 1956.
7. M. Morton, L. J. Fetters, R. A. Pett, and J. F. Meier, *Macromolecules* **3**, 327 (1970).
8. L. J. Fetters and M. Morton, *Macromolecules* **7**, 552 (1974).
9. T. Altares, D. P. Wyman, and V. R. Allen, *J. Polym. Sci., Part A* **2**, 4533 (1964).
10. M. Morton, E. E. Bostick, and R. G. Clarke, *J. Polym. Sci., Part A* **1**, 475 (1963).
11. L. J. Fetters, *J. Res. Natl. Bur. Stand., Sect. A* **69**, 33 (1965).
12. L. A. Utracki and J. E. L. Roovers, *Macromolecules* **6**, 366 (1973).
13. N. Hadjichristidis and J. E. L. Roovers, *J. Polym. Sci., Polym. Phys. Ed.* **12**, 2521 (1974).
14. W. W. Graessley and J. E. L. Roovers, *Macromolecules* **12**, 959 (1979).
15. A. F. Johnson and D. J. Worsfold, *J. Polym. Sci., Part A* **3**, 449 (1965).
16. M. Morton and F. R. Ells, *J. Polym. Sci.* **61**, 25 (1962).
17. H. S. Makowski and M. Lynn, *J. Macromol. Chem.* **1**, 443 (1966).
18. D. Margerison, D. M. Bishop, G. C. East, and P. McBride, *Trans. Faraday Soc.* **64**, 1872 (1968).
19. D. J. Worsfold, *J. Polym. Sci., Polym. Chem. Ed.* **5**, 2783 (1967).
20. D. J. Worsfold and S. Bywater, *Macromolecules* **5**, 393 (1972).
21. M. M. F. Al-Jarrah and R. N. Young, *Polymer* **21**, 119 (1980).
22. M. Morton, L. A. Falvo, and L. J. Fetters, *J. Polym. Sci., Part B* **10**, 561 (1972).

23. W. H. Glaze, J. E. Hanicak, M. L. Moore, and J. Chadhuri, *J. Organomet. Chem.* **44**, 39 (1972).
24. A. Hernandez, J. Semel, H. C. Broeker, H. G. Zachmann, and H. Sinn, *Makromol. Chem., Rapid Commun.* **1**, 75 (1980).
25. T. L. Brown, *J. Organomet. Chem.* **5**, 191 (1966).
26. G. Graham, S. Richtsmeier, and D. A. Dixon, *J. Am. Chem. Soc.* **102**, 5759 (1980).
26a. G. E. Coates, "Organometallic Compounds," p. 7. Wiley, New York, 1956.
27. R. Quirk and D. McFay, *J. Polym. Sci., Polym. Chem. Ed.* **19**, 1445 (1981).
28. J. E. L. Roovers and S. Bywater, *Macromolecules* **1**, 328 (1968).
29. F. Schue and S. Bywater, *Macromolecules* **2**, 458 (1969).
30. M. Morton, R. A. Pett, and L. J. Fetters, *Macromolecules* **3**, 333 (1970).
31. T. L. Brown, R. L. Gerteis, D. A. Bafus, and J. A. Ladd, *J. Am. Chem. Soc.* **84**, 1371 (1962).
32. T. L. Brown, J. A. Ladd, and G. Newman, *J. Organomet. Chem.* **3**, 1 (1965).

Chapter

6 Copolymerization

The classical treatment of the copolymerization of two monomers, developed several decades ago, is based on the four rate equations

$$M_1^* + M_1 \xrightarrow{k_{11}} M_1^* \tag{6.1}$$

$$M_1^* + M_2 \xrightarrow{k_{12}} M_2^* \tag{6.2}$$

$$M_2^* + M_2 \xrightarrow{k_{22}} M_2^* \tag{6.3}$$

$$M_2^* + M_1 \xrightarrow{k_{21}} M_1^* \tag{6.4}$$

where M_1 and M_2 represent the two types of monomers:

M_1^* = a growing chain having a terminal monomer unit M_1

M_2^* = a growing chain having a terminal monomer unit M_2

This approach is based on the assumption that only the *terminal* unit of a growing chain governs its reactivity. Although it was developed in connection with the free radical mechanism of addition polymerization, it is supposedly applicable to any other chain growth polymerization mechanism.

From the above four basic chemical equations, the classical kinetic expression was derived:

$$\frac{d[M_1]}{d[M_2]} = \frac{[M_1]}{[M_2]} \cdot \frac{r_1[M_1] + [M_2]}{r_2[M_2] + [M_1]} \tag{6.5}$$

where $r_1 = k_{11}/k_{12}$ and $r_2 = k_{22}/k_{21}$. Here r_1 and r_2 are known as the "reactivity ratios" of the two monomers, but actually represent the relative reactivity of each of the two different types of growing chains for its *own* type of monomer as opposed to the *other* monomer. In fact, these parameters can be considered as the "homopolymerization tendencies" of the two monomers. Equation (6.5) is, of course, an instantaneous expression, relating the composition of the copolymer, $d[M_1]/d[M_2]$, to the composition of the monomer mixture at *any given instant*. An integral equation, relating the composition of the cumulative copolymer at any conversion with the original monomer charge, has been derived as well.[1]

In applying Eq. (6.5) to anionic rather than radical polymerization, the following effects were considered important:

(i) The type of substituents present in the monomers should exert a marked effect on their reactivities toward the growing chains; i.e., electronegative substituents should render the double bond more electrophilic and therefore more reactive

(ii) Since the growing chains carry a negative charge, they should all exhibit similar relative reactivities toward the two monomers, regardless of resonance or steric considerations, i.e., $r_1 \sim 1/r_2$ or $r_1 r_2 \sim 1$.

Although these principles do apply in a number of cases, they cannot be considered universally true. This is so because, in ionic polymerization in general, and in the anionic case in particular, the "ionic" character of the growing chain end is markedly affected both by the nature of the counterion and by the medium. Thus the structure of the growing chain end can range all the way from a true dissociated carbanion to a covalent carbon–metal bond, depending on the type of cation and the solvent used. In other words, the r values in anionic copolymerization are not independent of the initiator used and the solvent, as in free radical systems, but have to be determined separately for each initiator–solvent combination.

1. Early Studies

The earlier work on copolymerization in anionic systems was carried out even before the mechanism was established. Thus Marvel, Bailey, and Inskeep,[2] in 1946, carried out the copolymerization of styrene and butadiene by sodium metal at a charge ratio of 75/25 (wt) butadiene/styrene (similar to that used for making the SBR type synthetic rubber). They showed that the copolymer contained 25 wt% styrene regardless of conversion and temperature, quite different from the emulsion polymerization (radical) system, which yields a lower initial styrene content (18%). Schulze and Crouch[3] also studied

TABLE 6.1 Anionic Copolymerization[a]

Monomer 1	Monomer 2	Initiator	Solvent	Remarks	References
Styrene	Methyl methacrylate	Na	Liq. NH_3	$r_1 = 0.12$; $r_2 = 6.4$	5
		n-C_4H_9Na	Benzene	MMA predominates	9
Styrene	Methyl methacrylate	Li	None	Equivalent reactivities	10, 11
		Li	H_4-furan	MMA predominates	10, 11
		n-C_4H_9Li	None	No detectable styrene in polymer	10, 11
		9-Fluorenyllithium (or Na)	Toluene or liq. NH_3 or H_4-furan	No detectable styrene in polymer	12
	Acrylonitrile	Li, Na, Be, Mg alkyls	None	$r_1 = 0.2$; $r_2 = 12.5$	13
	Vinyl acetate	Na	Liq. NH_3	$r_1 = 0.01$; $r_2 = 0.01$	5
	Butadiene	C_2H_5Na	Benzene	$r_1 = 0.96$; $r_2 = 1.6$	9
	Isoprene	Na, n-C_4H_9Na, Li	None or benzene	Isoprene predominates	14
			$(C_2H_5)_3N$ or $(C_2H_5)_2O$	Styrene predominates	14
		Na, n-C_4H_9Na, Li	H_4-furan	Styrene strongly predominates	14
	α-Methylstyrene	Alkyllithium	—	Molecular orbital calculation of effect of solvating agents (tetramethylethylenediamine and hexamethyltriphosphoramide) on copolymerization	15
	p-Methylstyrene	Na	H_4-furan	Styrene predominates	16
	or	Li	H_4-furan	Styrene predominates	16
	p-Methoxystyrene	n-C_4H_9Li	Toluene	Styrene predominates	16
	p-tert-Butylstyrene	sec-C_4H_9Li	Benzene or benzene/H_4-furan	$r_1 = 1.3$; $r_2 = 0.9$	16a
	p-Methylstyrene	sec-C_4H_9Li	Benzene or benzene/H_4-furan	$r_1 = 0.75$; $r_2 = 1.1$	
	p-Divinylbenzene	sec-C_4H_9Li	Benzene	Kinetics of gelation, p-DVB 10 times as reactive as isolated vinylbenzene	17

α-Methylstyrene	Butadiene	n-C_4H_9Na	Benzene	Diene predominates	9
	2,3-Dimethylbutadiene	n-C_4H_9Li	Benzene	Diene predominates	18
		n-C_4H_9Li	H_4-furan	Alternating copolymer	18
	Isoprene	C_2H_5Li	H_4-furan	$r_1 = 10$; $r_2 = 0.07$	19
p-Methylstyrene	*p*-Methoxystyrene	Na or Li	H_4-furan	*p*-Methyl predominates	16
		n-C_4H_9Li	Toluene	*p*-Methyl predominates	
Methyl methacrylate	Acrylonitrile	$NaNH_2$	Liq. NH_3	$r_1 = 0.25$; $r_2 = 7.9$	6
		Li, Na, Be, Mg alkyls	None	$r_1 = 0.34$; $r_2 = 6.7$	13
	Methyl acrylate	$NaNH_2$	Liq. NH_3	$r_1 = 0.1$; $r_2 = 4.5$	6
	Methacrylonitrile	Na	Liq. NH_3	$r_1 = 0.67$; $r_2 = 5.2$	7
	Vinyl acetate	$NaNH_2$	Liq. NH_3	$r_1 = 3.2$; $r_2 = 0.4$	6
	Cinnamyl methacrylate	n-C_4H_9Li	Toluene (−78°C)	$r_1 = 0.80$; $r_2 = 1.15$	20
Acrylonitrile	Vinyl butyl sulfone	Na	Liq. NH_3	$r_1 = 1.1$; $r_2 = 0.2$	8
	Methyl acrylate	n-C_4H_9Li	Toluene	Composition studies	21
	Nitropropene	n-C_4H_9Li or CH_3ONa	–	$r_1 = 0.03$; $r_2 = 8.1$	22
Acrolein	Acrylonitrile	Na	–	Transparent copolymer	23
	Vinyls	Na or Li	Acetone	No copolymerization with styrene or methyl methacrylate, but copzn. with acrylonitrile	24
	Epoxides	NaCN	Dimethylformamide/H_4-furan	Transparent film	25
Aldehydes	Isocyanates	n-C_4H_9Li, KCN, Na naph	Diethyl ether, H_4-furan, *n*-hexane, dimethylformamide	Aldehydes predominate in nonpolar solvents, but less in polar solvents	26
n-Butyaldehyde	*n*-Butyl isocyanate	n-C_4H_9Li	*n*-Hexane	$r_1 = \infty$; $r_2 = 0$	26
		n-C_4H_9Li	Diethyl ether	$r_1 = 500$; $r_2 = 10$	26
		n-C_4H_9Li	H_4-furan	$r_1 = 70$; $r_2 = 10$	26

(*table continues*)

TABLE 6.1 Anionic Copolymerization (*continued*)

Monomer 1	Monomer 2	Initiator	Solvent	Remarks	References
Aldehydes (*continued*)					
Chloral	*n*-Butyl isocyanate	$n\text{-}C_4H_9Li$	*n*-Hexane	$r_1 = \infty$; $r_2 = 0$	26
		$n\text{-}C_4H_9Li$	Diethyl ether	$r_1 = 2\text{–}50$; $r_2 = 0$	26
	Phenyl isocyanate	$n\text{-}C_4H_9Li$	*n*-Hexane	$r_1 = \infty$; $r_2 = 0$	26
		$n\text{-}C_4H_9Li$	Diethyl ether	$r_1 = 0.5$; $r_2 = 0$	26
		$n\text{-}C_4H_9Li$	Benzene	Crystalline copolymer	27
		$(C_2H_5)_3N$	–	Amorphous copolymer	27
β-Cyanopropionaldehyde	Phenyl isocyanate	Benzophenone diK or diNa	H_4-furan	No reasonable *r* values, some heterogeneity	28
		Na naph, Na benzylate	H_4-furan	No reasonable *r* values, some heterogeneity	28
		Benzophenone diLi	H_4-furan	$r_1 = 0.18$; $r_2 = 0.8$	28a
Anisaldehyde	Dimethylketene	Benzophenone diK or diNa	H_4-furan	Ketene predominates	29
		Benzophenone diLi	H_4-furan	Alternating copolymer	29
2,4-Toluene diisocyanate	Vinyl monomers (styrene, methyl methacrylate, acrylonitrile)	1,4-Diazabicyclo-[2.2.2]octane	None	$r_1 = 0.13\text{–}0.17$; $r_2 \sim 0$	30
Diphenylketene	Ethyl isocyanate	$n\text{-}C_4H_9Li$	Toluene	Crystalline copolymer (T_m 200°C)	31
	Phenyl isocyanate	NaCN	Dimethylformamide	Crystalline copolymer (T_m 200°C)	31

[a] Data on organolithium copolymerization of styrene and dienes are listed separately in Table 6.2.

the sodium polymerization of these two monomers and also reached the same conclusion as the previous authors; i.e., that this system does not proceed by a free radical mechanism. As additional evidence they also included some data on the chain unit structure of the butadiene in the copolymer, which exhibited a much higher content of the 1,2-butadiene units than the emulsion copolymer.

Mayo and co-workers[4] (1950) used the copolymerization of several pairs of monomers as a diagnostic tool to distinguish between the three polymerization mechanisms: free radical (peroxides), cationic (Lewis acids), and anionic (sodium metal). They obtained clear-cut evidence that these three types of initiators operated by different mechanisms since they gave markedly different results. Thus, in the case of the sodium polymerizations, (i) methyl methacrylate polymerized almost exclusively in the presence of styrene, (ii) acrylonitrile was very strongly preferred in the presence of methyl methacrylate, and (iii) a styrene–diethyl fumarate mixture showed no polymerization at all.

Landler[5,6] and Foster[7,8] were among the main contributors of information about the few anionic copolymerization systems studied at that time, using as initiators sodium metal in liquid ammonia as solvent. These studies were complicated by the fact that the systems used were heterogeneous, involving surface reactions. One homogeneous system studied used sodium amide in liquid ammonia. Only a few monomers were used in these studies, including styrene, methyl methacrylate, methyl acrylate, acrylonitrile, methacrylonitrile, vinyl acetate, and vinyl butyl sulfone. The results are included in the listing shown in Table 6.1 and can be summarized by the statement that most of the r values found indicated a very strong preference for one of the monomers in each case, undoubtedly due to the much greater polarity of the preferred monomer.

2. Later Work

Much more information became available about anionic copolymerization after the advent of the Ziegler–Natta organometallic catalysts and the renewed interest in the alkali organometallic compounds during the late 1950s. The latter initiators can be used in homogeneous polymerization systems with such important monomers as styrene, butadiene, isoprene, and methyl methacrylate in various solvents, both polar and nonpolar. The alkali metal alkyls, and specifically the lithium alkyls, are especially desirable as initiators, since they can be used in homogeneous solution in a variety of solvents. Furthermore, since these systems generally show no signs of chain termination reactions, under suitable conditions they are amenable to calculations of actual concentrations of growing chains.

A listing of available data on anionic copolymerizations is shown in Tables 6.1 and 6.2 for the reader's convenience. Most of these data are consistent, although there may be some contradictions. Because of its special relevance to synthetic rubber technology, and because of the extensive data available, the copolymerization of the 1,3-dienes with styrene is described separately in Table 6.2 and will be discussed later. There are, however, certain features of the data in Table 6.1 that deserve special discussion.

TABLE 6.2 Organolithium Copolymerization of Dienes and Styrene

Monomer 1	Monomer 2	Initiator	Solvent	Temp. (°C)	Reactivity ratios r_1	r_2	References
Styrene	Butadiene	n-C_4H_9Li	Benzene	30–50	0.035	10.0	9
					0.05	20	32
		C_2H_5Li	Toluene	25	0.1	12.5	33
		n-C_4H_9Li	n-Heptane	30	0.1	7	34
			Benzene	29	0.3	4.5	35, 36
			Cyclohexane	40	0.04	26	37
			n-Hexane	50	0.025	15.1	38
			$C_2H_5OC_2H_5$	30	0.11	1.78	39
		C_2H_5Li	$(C_2H_5)_3N$ (40% in toluene)	25	0.3	5.5	33
			H_4-furan	−35	0.2	8	33
	Isoprene	n-C_4H_9Li	None	25	*a*	*a*	14
			Benzene	25	*a*	*a*	14
				30	0.14	7.0	40
				30	0.26	10.6	36
		C_2H_5Li	Toluene	27	0.25	9.5	41
		n-C_4H_9Li	Cyclohexane	40	0.046	16.6	42
			H_4-furan	25	*b*	*b*	14
		C_2H_5Li	H_4-furan	27	9.0	0.1	33
				−35	40	0	33
		Li or	$C_2H_5OC_2H_5$	25	*b*	*b*	14
		n-C_4H_9Li	$(C_2H_5)_3N$	25	*b*	*b*	14
		C_2H_5Li	$(C_2H_5)_3N$	27	0.8	1.0	41
Butadiene	Isoprene	n-C_4H_9Li	n-hexane	50	3.38	0.47	41a
			Benzene	40	3.6	0.5	43
			H_4-furan	20	4.5	0.13	44
			$(C_2H_5)_3N$	20	3.6	0.11	44

[a] r values not given, but isoprene predominates.
[b] r values not given, but styrene predominates.

Tobolsky *et al.*[10,11] showed that different types of copolymers were obtained when metallic lithium was used instead of the soluble n-butyllithium in the copolymerization of styrene and methyl methacrylate. Thus the former seemed to lead to formation of a block copolymer consisting of a block of poly(methyl methacrylate) attached to a block of a styrene–methyl methacrylate copolymer, while the latter led to a poly(methyl methacrylate) homopolymer. This was explained by postulating the simultaneous existence of two mechanisms, i.e., free radical and anionic, the former due to formation of free radicals during the electron-transfer initiation step and leading to a random copolymer, while the latter, of course, leads to the exclusive anionic polymerization of the methyl methacrylate. (The inability of the methacrylate anion to attack styrene monomer was also very elegantly confirmed by the work of Goode *et al.*[12])

However, later work by Overberger and Yamamoto,[45] using NMR spectroscopy, showed that the above copolymer actually consisted of a *diblock* of polystyrene and poly(methyl methacrylate), which could not therefore involve the random copolymerization of a free radical mechanism. They therefore proposed, with some supporting evidence, that the polystyrene block was first initiated at the surface of the lithium metal and, after it attained a certain size, became detached from the surface and initiated the homogeneous anionic polymerization of the methyl methacrylate. This theory also had some support from the work of Korotkov,[46] who also obtained unusual block copolymers in the copolymerization of these two monomers by methyllithium, which is *insoluble* and hence involves a heterogeneous initiation reaction but not electron transfer as in the case of the lithium metal. Hence no free radical intermediate can be invoked in that case.

The work of Zutty and Welch[13] on the copolymerization of styrene and acrylonitrile, and methyl methacrylate and acrylonitrile, also deserves some discussion. The data in Table 6.1 show that there is a strong preference for acrylonitrile, presumably due to the greater polarity and electrophilicity of its double bond. However, when the alkyls of zinc and cadmium were used in these systems, a more random copolymer was obtained, corresponding to that usually resulting from free radical polymerization. Hence they concluded that the organometallics of groups I, IIA, and IIIB (B and Al) all lead to anionic polymerization, whereas initiators based on metals of group IIB somehow work by a free radical mechanism.

It should also be pointed out that the data in Tables 6.1 and 6.2 were generally obtained by the usual method of determining the composition of the polymer (at low conversion) resulting from a known charge ratio of the monomers. However, this is not always necessary in the case of anionic copolymerization, where termination of growing chains does not occur. As a matter of fact, the usual procedure may give misleading results if there is a substantial competition between the initiation and propagation steps, as often happens in anionic systems. Furthermore, it may actually be possible in these termination-free anionic systems to measure *directly* the individual propagation rate constants and thus to calculate the r values *ab initio*.

This type of determination was actually carried out in the case of the organolithium copolymerization of styrene with butadiene or isoprene[35,36] (see Table 6.2). The r values in this case were calculated directly from the four propagation rate constants, the cross-propagation rate constants (k_{12} and k_{21}) being obtained by measuring the rate of reaction between polystyryllithium and the diene on the one hand, and polydienyllithium and styrene on the other. Since the ultraviolet absorbance of polystyryllithium and polydienyllithium species differed substantially, it was possible to use it as an analytical method for the rate of appearance or disappearance of the polystyryllithium chain ends.

3. Copolymerization of Styrene and 1,3-Dienes

The anionic copolymerization of the two dienes, butadiene and isoprene, with styrene has been studied most extensively in the homogeneous organolithium systems and the data obtained have been collated in Tables 6.2, 6.3, and 6.4. The most striking feature of these systems is, of course, the effect of polar solvents. Thus in the case of nonpolar media, either with undiluted monomer or in hydrocarbon solvents, there is an overwhelming preference for the diene, and the styrene monomer is almost excluded, despite the fact that styrene alone shows a much faster polymerization rate than either of the dienes. However, the addition of a polar solvent, as exemplified by the ethers, has a marked effect

TABLE 6.3 Effect of Solvents on Organolithium Copolymerization of Butadiene and Styrene[a]

Solvent	Temp. (°C)	Reactivity ratios	
		r_B	r_S
None	25	11.2 ± 0.1	0.04 ± 0.01
Benzene	25	10.8 ± 0.2	0.04 ± 0.01
n-Hexane	0	13.3	0.03
	25	12.5 ± 0.2	0.03 ± 0.01
	50	11.8	0.04
Cyclohexane	25	15.5 ± 0.3	0.04 ± 0.02
H_4-furan	−78	0.04 ± 0.01	11.0 ± 0.1
	0	0.2	5.3
	25	0.3	4.0
Diethyl ether	25	1.7 ± 0.08	0.4 ± 0.05
Triethylamine	25	3.5 ± 0.13	0.5 ± 0.06
Anisole	25	3.4 ± 0.05	0.3 ± 0.02
Diphenyl ether	25	2.8 ± 0.11	0.1 ± 0.05

[a] Initiator — *sec*-C_4H_9Li. From Morton and Huang.[17]

TABLE 6.4 Effect of Polar Cosolvents on Organolithium Copolymerization of Butadiene–Styrene in Nonpolar Media[a]

Vol % cosolvent	Reactivity ratios							
	Hexane/ H_4-furan		Benzene/ H_4-furan		Benzene/ $C_2H_5OC_2H_5$		Benzene/ $(C_2H_5)_3N$	
	r_B	r_S	r_B	r_S	r_B	r_S	r_B	r_S
None	12.5	0.03	10.8	0.04	10.8	0.04	10.8	0.04
1	1.7	0.6	1.7	0.7	2.7	0.1	5.0	0.1
10	0.8	1.5	0.8	1.6	2.4	0.1	4.8	0.3
50	0.5	2.7	0.5	2.4	2.0	0.2	—	—
100	0.3	4.0	0.3	4.0	1.7	0.4	3.5	0.5

[a] From Morton and Huang.[47]

in "reversing" this behavior, favoring the inclusion of styrene in the copolymer. This effect is especially dramatic in the case of such highly solvating solvents as H_4-furan, even a 1% (or smaller) addition being sufficient to reverse the r values.

The copolymerization behavior of butadiene/styrene under various conditions has been the subject of a detailed recent study,[47] using rigorous, high vacuum techniques to avoid the presence of adventitious impurities. The results are summarized in Tables 6.3 and 6.4, which probably represent the most accurate r values obtained for these monomers. The variation between the effects of the different polar solvents is presumably a reflection of their solvating power for the lithium cation. Especially interesting is the effect of temperature, which is negligible in the case of nonpolar media (n-hexane) while showing a substantial effect in the case of the polar solvent H_4-furan. These results are easily rationalized. As a rule, r values are not markedly influenced by temperature, since they represent the ratio of the two types of propagation steps which do not differ much in activation energy. However, the temperature can be expected to affect substantially any solvation phenomenon, and this is reflected in the data on H_4-furan (Table 6.3). Thus, as the temperature is *increased* from $-78°C$ to $25°C$, the extent of solvation would be expected to decrease, making the carbon–lithium bond less ionic and more covalent, as in nonpolar media.

The effect of polar solvents on the organolithium copolymerization of styrene and the 1,3-dienes described above has been generally explained on the basis of the changes in the carbon–lithium bond structure caused by such solvents. These are discussed in detail in Chapter 7, based on convincing evidence from nuclear magnetic resonance studies which shows that polar solvents bring about a delocalization of the carbon–lithium bond of the chain end. Hence polar solvents also favor a 1,2 (or 3,4) monomer addition of the 1,3-dienes (see Table 7.3).

As for the case of nonpolar media, the virtual exclusion of styrene in the copolymerization has not yet been entirely rationalized. It is known that, in such media, the carbon–lithium bond of the chain is mainly covalent (see Chapter 7), and that these covalent chain ends are strongly associated in pairs (see Chapter 5). Hence it has been proposed that monomer addition involves a concerted four-center reaction between the diene and the associated chain ends (Fig. 7.7), thus making it difficult for styrene to be included.* The addition of a polar solvent is known to delocalize the carbon–lithium bond and to break up the chain-end association, thus facilitating a more "ionic" reaction.

As mentioned previously, the organolithium copolymerization of butadiene and styrene is of considerable importance in synthetic rubber technology.[49] Such copolymers require a random placement of styrene units (~ 25%) as well as a high content of 1,4-butadiene units, conditions which appear to be mutually exclusive. Hence two approaches have been used to circumvent this difficulty: (i) the use of a weak polar "modifier"[50] which can help to include the required amount of random styrene units without raising the vinyl content of the butadiene units above 30% or (ii) the use of a continuous polymerization system (in a hydrocarbon medium) which maintains a high styrene/butadiene ratio in the reactor and a low conversion per pass.

References

1. F. R. Mayo and F. M. Lewis, *J. Am. Chem. Soc.* **66**, 1594 (1944).
2. C. S. Marvel, W. J. Bailey, and G. E. Inskeep, *J. Polym. Sci.* **1**, 275 (1946).
3. W. A. Schulze and W. W. Crouch, *J. Am. Chem. Soc.* **70**, 3891 (1948).
4. C. Walling, E. R. Briggs, W. Cummings, and F. R. Mayo, *J. Am. Chem. Soc.* **72**, 48 (1950).
5. I. Landler, *Proc. Int. Colloq. Macromol., 1949* p. 377 (1950); *C. R. Hebd. Seances Acad. Sci.* **230**, 539 (1950).
6. I. Landler, *J. Polym. Sci.* **8**, 63 (1952).
7. F. C. Foster, *J. Am. Chem. Soc.* **72**, 1370 (1950).
8. F. C. Foster, *J. Am. Chem. Soc.* **74**, 2299 (1952).
9. A. A. Korotkov, *Int. Symp. Macromol. Chem., 1957* Doc. II, Pap. No. 60 (1957); *Angew. Chem.* **70**, 85 (1958).
10. K. F. O'Driscoll, R. J. Boudreau, and A. V. Tobolsky, *J. Polym. Sci.* **31**, 115 (1958).
11. K. F. O'Driscoll and A. V. Tobolsky, *J. Polym. Sci.* **31**, 123 (1958); **37**, 363 (1959).
12. R. K. Graham, D. L. Dunkelberger, and W. E. Goode, *J. Am. Chem. Soc.* **82**, 400 (1960).
13. N. L. Zutty and F. J. Welch, *J. Polym. Sci.* **43**, 445 (1960).

* In this connection, it is interesting to note the work of Smart *et al.*,[48] who obtained direct evidence for an interaction between the double bond in 3-butenyllithium (in hydrocarbon media) and the lithium atom, using ^{7}Li NMR, UV, and IR spectroscopy.

14. D. J. Kelley and A. V. Tobolsky, *J. Am. Chem. Soc.* **81**, 1597 (1959).
15. W. Kampf and H. Weber, *Angew. Makromol. Chem.* **31**, 25 (1973).
16. A. V. Tobolsky and R. J. Boudreau, *J. Polym. Sci.* **51**, 556 (1961).
16a. J. Chen and L. J. Fetters, *Polym. Bull.* **4**, 275 (1981).
17. D. J. Worsfold, *Macromolecules* **3**, 514 (1970).
18. Y. Heimi, Y. Okamoto, H. Takano, and K. Ohta, *Polym. J.* **3**, 442 (1972).
19. D. K. Polyakov, A. R. Gantmakher, and S. S. Medvedev, *Vysokomol. Soedin., Ser. A* **9**, 1390 (1967); *CA* **67**, 82811k.
20. K. Kawai and T. Ichihashi, Japanese Patent 72/33,074 (1972); *CA* **78**, 1127926.
21. I. L. Artamonova, S. T. Klenin, A. V. Troitskaya, and B. L. Erussalimsky, *Makromol. Chem.* **175**, 2329 (1974).
22. E. G. Gorlov and V. P. Alaniya, *Izv. Vyssh. Uchebn. Zaved. Khim. Khim. Tekhnol.* **17**(10), 1601 (1974); *CA* **82**, 58258 (1975).
23. J. Mateo Lopez and R. Sastre Muñoz, Spanish Patent 403,715 (1975); *CA* **84**, P7487.
24. J. Mateo Lopez and R. Sastre Muñoz, *Rev. Plast. Mod.* **21**, 454 (1970).
25. J. Mateo Lopez and R. Sastre Muñoz, Spanish Patent 410,773 (1975); *CA* **85**, P22104.
26. G. Odian and L. S. Kiraoka, *J. Macromol. Sci., Chem.* **6**, 109 (1972).
27. J. Mateo Lopez, R. Sastre Muñoz, and M. Garcia Perez, Spanish Patent 413,085 (1976); *CA* **85**, P63660r.
28. K. Hashimoto and H. Sumitomo, *J. Polym. Sci., Polym. Chem. Ed.* **12**, 1569 (1974).
28a. K. Hashimoto and H. Sumitomo, *J. Polym. Sci., Polym. Chem. Ed.* **9**, 1747 (1971).
29. K. Hashimoto, H. Sumitomo, and N. Hayakawa, *Polym. J.* **6**, 391 (1974).
30. V. N. Nikolaev, A. V. Kostochko, A. I. Petrov, R. A. Lasarev, M. M. Izheeva, A. A. Matveeva, and V. M. Lukashin, *Khim. Khim. Tekhnol.* (Cheboksary) **2**, 17 (1971); *CA* **78**, 44052h (1973).
31. H. Higashi and H. Harada, Japanese Patent 69/11,140 (1969); *CA* **71**, P71176p.
32. A. A. Korotkov and N. N. Chesnokova, *Polym. Sci. USSR* (*Engl. Transl.*) **2**, 284 (1961).
33. Yu. L. Spirin, A. A. Arest-Yakubovich, D. K. Polyakov, A. R. Gantmakher, and S. S. Medvedev, *J. Polym. Sci.* **58**, 1161 (1962).
34. I. Kuntz, *J. Polym. Sci.* **54**, 569 (1961).
35. M. Morton and F. R. Ells, *J. Polym. Sci.* **61**, 25 (1962).
36. F. R. Ells, Ph.D. Dissertation, University of Akron, Akron, Ohio, 1963.
37. A. F. Johnson and D. J. Worsfold, *Makromol. Chem.* **85**, 273 (1965).
38. V. D. Mochel, *Rubber Chem. Technol.* **40**, 1200 (1967).
39. A. A. Korotkov, S. P. Mitzengendler, and K. M. Aleyev, *Polym. Sci. USSR* (*Engl. Transl.*) **3**, 487 (1962).
40. G. A. Rakova and A. A. Korotkov, *Polym. Sci. USSR* (*Engl. Transl.*) **3**, 990 (1962).
41. Yu. L. Spirin, D. K. Polyakov, A. R. Gantmakher, and S. S. Medvedev, *J. Polym. Sci.* **53**, 233 (1961).
41a. G. V. Rakova and A. A. Korotkov, *Rubber Chem. Technol.* **33**, 623 (1960).
42. D. J. Worsfold, *J. Polym. Sci., Polym. Chem. Ed.* **5**, 2783 (1967).
43. J. Furukawa, T. Saegusa, and K. Irako, *Kagaku Zasshi* **65**, 2029 (1962); *CA* **58**, 11553 (1963).
44. L. S. Bresler, B. A. Dolgoplosk, M. F. Kolechkova, and E. N. Kropacheva, *Proc. Acad. Sci. USSR* **144**, 420 (1963).
45. C. G. Overberger and N. Yamamoto, *J. Polym. Sci., Part B* **3**, 569 (1965); *J. Polym. Sci., Polym. Chem. Ed.* **4**, 310 (1966).
46. A. A. Korotkov, private communication.
47. M. Morton and L.-K. Huang, unpublished data; L.-K. Huang, Ph.D. Dissertation, University of Akron, Akron, Ohio, 1979.
48. J. P. Oliver, J. B. Smart, and M. T. Emerson, *J. Am. Chem. Soc.* **88**, 4101 (1966).
49. R. S. Hanmer and H. E. Railsbach, *Rubber Age* **96**, 73 (1964).
50. H. L. Hsieh, *J. Polym. Sci., Polym. Chem. Ed.* **8**, 533 (1965).

Chapter

7 Organolithium Polymerization of Dienes

The first case of anionic polymerization of 1,3-dienes probably arose out of the quest for synthetic rubber, as illustrated by the work of Matthews and Strange in 1910[1] and of Harries in 1911[2,3] on the polymerization of isoprene by means of sodium or potassium. This discovery then led to the development of the synthetic rubber industry in Germany and the USSR, based on the sodium polymerization of butadiene. Although the process resulted in the production of an acceptable synthetic rubber, little was known of the isomeric chain unit structure other than the fact that the 1,2-units predominated, since such information could be obtained by ozonolysis of the polymer. It was only much later, with the advent of infrared spectroscopy, that more detail became available about the chain microstructure. Thus Meyer *et al.*[4] showed that potassium led to a somewhat different microstructure in polybutadiene (less 1,2) than sodium, illustrating the effect of the counterion in these anionic polymerizations.

The first unambiguous report of the use of lithium in diene polymerization appears to be that of Ziegler and co-workers,[5] who studied the reaction between the alkali metals lithium and sodium as well as alkyllithium with butadiene, isoprene, and 2,3-dimethylbutadiene. However, again little was known about the details of the isomeric chain-unit structure until more sophisticated analytical methods became available. It is not surprising, therefore, that it was during the early 1950s, shortly after infrared spectroscopy was developed, that the startling discovery was made[6] that lithium is capable of polymerizing isoprene to a very high cis-1,4 chain unit content, close to that of natural rubber (Hevea). This discovery aroused a great deal of interest in the factors affecting the isomeric chain unit structure in the polymerization of 1,3-dienes by alkali metals, and especially in the use of lithium and its organic derivatives.

1. Chain Microstructure: Effect of Initiator and Solvents

The intensive investigations which followed the discovery of the virtually stereospecific polymerization of isoprene by lithium soon showed that it was not only the type of alkali metal but the nature of the solvent which exerts a profound effect on the chain unit structure[7-13] of polydienes. Fairly

TABLE 7.1 Chain Structure in Alkali-Metal-Initiated Polymerization of Butadiene and Isoprene (by Infrared Spectroscopy)

		Chain microstructure (mole %)				
Initiator	Solvent	*cis*-1,4	*trans*-1,4	1,2	3,4	References
		Polyisoprene				
$n\text{-}C_4H_9Li$	*n*-Heptane	93	0	0	7	9
$n\text{-}C_4H_9Li$	H_4-furan	0	30	16	54	9
C_6H_5Na	*n*-Heptane	0	47	8	45	9
C_6H_5Na	H_4-furan	0	38	13	49	9
$C_6H_5CH_2K$	*n*-Heptane	0	52	10	38	9
$C_6H_5CH_2K$	H_4-furan	0	43	17	40	9
Rb	None	5	47	10	38	7
Cs	None	4	51	8	37	7
$C_{10}H_8Cs^a$	H_4-furan	0	28	29	43	15
		Polybutadiene				
C_2H_5Li	*n*-Hexane	43	50	7		15
C_2H_5Li	H_4-furan	0	9	91		15
Na	None	10	25	65		7
$C_{10}H_8Na^a$	H_4-furan	0	9	91		15
K	None	15	40	45		7
$C_{10}H_8K^a$	H_4-furan	0	17	83		15
Rb	None	7	31	62		7
Rb	H_4-furan	0	25	75		15
Cs	None	6	35	59		7
$C_{10}H_8Cs^a$	H_4-furan	0	14	86		15

[a] −78°C. All others at 25–50°C.

comprehensive reviews of these findings have also been published.[14,15] Typical data on the effect of various alkali metals (and organometallics) and solvents on the microstructure of polyisoprene and polybutadiene are shown in Table 7.1. These are older data and are based mainly on infrared analysis, which is not unambiguous in distinguishing between cis- and trans-1,4 structures in polyisoprene. More recent data from NMR spectroscopy are more reliable in that respect. However, Table 7.1 is useful in showing the relative effects of counterions and solvents.

The conclusions reached from these extensive investigations can be summarized as follows: (i) the proportion of 1,4 chain units is highest for lithium and decreases with increasing electropositive character of the alkali metal, (ii) polar solvents (e.g., ethers or amines) lead to a higher side-vinyl structure in the polymer, especially in the case of lithium. The formation of these side-vinyl units (1,2 or 3,4) has thus been related to the ionic character of the carbon–metal bond at the tip of the growing chain, polar solvents apparently increasing this ionicity by solvating the metal cation. As noted in a previous chapter, the ionic character of the growing chain also strongly affects copolymerization behavior. Tobolsky and Rogers[9] have actually correlated the chain structure of the isoprene units to the copolymerization of styrene and isoprene, as shown in Table 7.2. It is obvious that those systems which make it possible to include a reasonable proportion of styrene in the copolymer also result in a lower 1,4 content, which incidentally poses problems in synthetic rubber technology.

All of the above results are consistent with the kinetic studies discussed previously, where the propagating chain ends in polar solvents were considered to be in the form of ion pairs in equilibrium with free carbanions, while in

TABLE 7.2 Chain Structure in Isoprene–Styrene Copolymerization[a] (Equimolar Styrene–Isoprene Charge)

Initiator	Solvent	% 1,4 Units[b]	% Styrene[c]
Li or n-C_4H_9Li	Hydrocarbons	95	16
n-C_4H_9Li	Diphenyl ether	82	30
n-C_4H_9Li	Dibutyl sulfide	62	50
Li	Triethylamine	52	60
Na	Hydrocarbons	45	62
Li or n-C_4H_9Li	Diethyl ether	51	68
n-C_4H_9Li	Dimethyl sulfide	36	75
Na	Dimethyl ether	36	75
Na	H_4-furan	32	80
Li or n-C_4H_9Li	H_4-furan	30	80

[a] From Tobolsky and Rogers.[9]
[b] By infrared spectroscopy.
[c] By weight, in initial copolymer.

hydrocarbon media they were largely nonionic in character. As regards the stereospecificity of the propagation step, the most thoroughly investigated systems have been those involving lithium and organolithium initiators. This is because the latter are the most versatile type, being soluble in a variety of solvents, and can be controlled to undergo relatively few side reactions (termination or transfer). Hence it is not surprising that these systems have provided the most fertile ground for studies of chain structure and its relation to reaction mechanism.

As previously mentioned, the most interesting and reliable data on chain microstructure of the polydienes have been obtained in more recent years by NMR spectroscopy, using both ^{1}H and ^{13}C resonance. A tabulation of the more interesting data is compiled in Tables 7.3 to 7.5 for polyisoprene and polybutadiene. These data have been selected to illustrate the effects of (i) solvents, (ii) initiator concentration, and (iii) temperature on the chain microstructure. (The *type* of organolithium initiator cannot be expected to have any effect, since the organic moiety is not associated with the growing chain-end.)

a. Effect of Solvents

From the data in Table 7.3 it can be readily seen that polar solvents, especially H_4-furan, can drastically alter the polyisoprene structure from the 70–90% cis-1,4 obtained in hydrocarbon media to mainly 3,4 (and some 1,2). A similar effect occurs with polybutadiene, which changes from a very high ($> 90\%$) 1,4 content in hydrocarbons to a very high 1,2 content in H_4-furan. These drastic changes generally occur even when very low concentrations of such polar cosolvents as H_4-furan are added to hydrocarbon media. Other ether or amine cosolvents may not have such a dramatic effect[8]. The cis–trans ratio for polybutadiene prepared in hydrocarbon media is generally about 2:3, and this ratio appears to be maintained even at high vinyl contents.

As previously stated, this effect of polar solvents is presumably a natural outcome of the solvation of the lithium cation, leading to a more ionic character of the carbon–lithium bond. What is perhaps unexpected is the more subtle effect of *different* nonpolar media (hydrocarbons) on the chain microstructure. This is exemplified in Table 7.3 by the noticeable difference in cis–trans ratio of polyisoprene produced in benzene versus that in *n*-hexane or cyclohexane, even though the 3,4 content remains virtually unchanged. It may be noted at this point, incidentally, that no evidence could be found for any 1,2 chain units in polyisoprene prepared in nonpolar media, within the limits of detection of the instrumentation (300 MHz ^{1}H NMR),[17] i.e., approximately 1%.

TABLE 7.3 Effect of Solvents on Chain Microstructure in Alkyllithium-Initiated Polymerization of Isoprene and Butadiene[a]

Polymer	Solvent system: Nonpolar	Solvent system: Polar	[I] (× 10³)	[S]/[I][b]	Microstructure (mole %)[c]: *cis*-1,4	*trans*-1,4	3,4	1,2	References
Polyisoprene	Benzene	—	0.3	—	69	25	6	—	16, 17
	Benzene	—	0.3	—	71	23	6	—	18
	Cyclohexane	—	0.3	—	80	15	5	—	19
	n-Hexane	—	0.03	—	81	15	4	—	16, 17
	n-Hexane	H_4-furan	0.3	2	68	19	13	—	19
	n-Hexane	H_4-furan	0.3	15	—	69	31	—	19
	n-Hexane	H_4-furan	0.3	*d*	—	26	66	9	19
Polybutadiene	*n*-Hexane	—	0.2	—	35	57	—	8	16, 20, 21
	n-Hexane	Diethyl ether	0.2	6.3	29	50	—	21	16, 21
	n-Hexane	H_4-furan	0.2	1.0	25	40	—	35	16, 21
	n-Hexane	H_4-furan	0.2	8.2	21	31	—	45	16, 21
	n-Hexane	H_4-furan	0.2	17	14	28	—	58	16, 21
	n-Hexane	H_4-furan	0.2	53	7	10	—	85	16, 21

[a] 0.5M Monomer, 20–30°C.
[b] [S]/[I] = molar ratio of polar solvent to initiator.
[c] Precision of ^{1}H NMR data: *cis*-1,4 or *trans*-1,4 = ± 2%; 3,4 or 1,2 = ± 1%.
[d] 10 vol% H_4-furan.

b. Effect of Initiator Concentration (Nonpolar Media)

The effects of different hydrocarbon solvents as well as of the concentration of initiator are shown in Tables 7.4 and 7.5. The following conclusions can be drawn from these results:

(i) Aliphatic solvents, e.g., hexane and cyclohexane, lead to a slightly higher cis-1,4 content than aromatic solvents (benzene), without much effect on side-vinyl content

(ii) The absence of any solvents (undiluted monomer) leads to the highest attainable cis-1,4 content for both polyisoprene and polybutadiene. This appears to be a function of monomer concentration[16,17]

(iii) A decrease in initiator concentration leads to a noticeable increase in cis-1,4 content for both polymers, without any notable change in side-vinyl content. Hence the polymerization of isoprene and butadiene in the undiluted state, at initiator concentrations of about 10^{-5} *M* results in a 96% cis-1,4 polyisoprene and an 86% cis-1,4 polybutadiene.

The combined effects of both initiator and monomer concentration on cis-1,4 content for both of these dienes is shown graphically in Figs. 7.1 and 7.2.

TABLE 7.4 Effect of Initiator and Monomer Concentration on Chain Microstructure in Alkyllithium-Initiated Polymerization of Isoprene in Nonpolar Media[a]

			Microstructure (mole %)		
Solvent	[I]	[M]	*cis*-1,4	*trans*-1,4	3,4
Benzene	9×10^{-3}	0.5	69	25	6
Benzene	4×10^{-5}	0.5	70	24	6
n-Hexane	1×10^{-2}	0.5	70	25	5
n-Hexane	1×10^{-5}	0.5	86	11	3
None	3×10^{-3}	10	77	18	5
None	8×10^{-6}	10	96	0	4
None[b]	1×10^{-5}	10	95	1	4

[a] At 20°C. From Morton and Fetters,[16] Rupert,[17] and Santee *et al.*[21]
[b] 46°C.

TABLE 7.5 Effect of Initiator and Monomer Concentration on Chain Microstructure in Alkyllithium-Initiated Polymerization of Butadiene in Nonpolar Media[a]

			Microstructure (mole %)		
Solvent	[I]	[M]	*cis*-1,4	*trans*-1,4	1,2
Benzene	8×10^{-6}	0.5	52	36	12
Cyclohexane	1×10^{-5}	0.5	68	28	4
n-Hexane	2×10^{-5}	0.5	56	37	7
n-Hexane	2×10^{-4}	0.5	34	58	8
n-Hexane	3×10^{-2}	0.5	30	62	8
None	3×10^{-3}	12	39	52	9
None	4×10^{-5}	12	71	22	7
None	7×10^{-6}	12	86	9	5
[b]None	7×10^{-6}	12	86	9	5

[a] At 20°C. From Morton and Fetters,[16] Rupert,[17] and Santee *et al.*[21]
[b] 0°C.

Since the initial monomer concentration appears to have a marked effect on the cis–trans ratio, it might be expected that this ratio would show a gradual change during the course of the polymerization. However, an examination[17] of the chain microstructure at different stages in the polymerization of undiluted isoprene at 20°C showed no noticeable changes over a range of conversion from 13% to 86%.

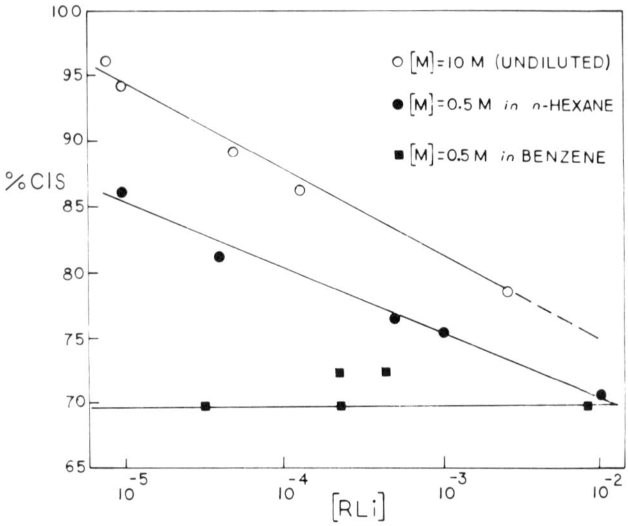

Fig. 7.1 Effect of nonpolar solvents and organolithium concentration on polyisoprene chain structure.[17]

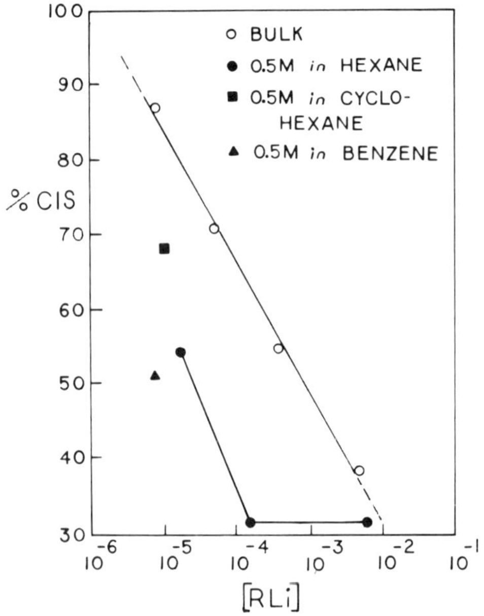

Fig. 7.2 Effect of nonpolar solvents and organolithium concentration on polybutadiene chain structure.[17]

c. Effect of Temperature

The temperature of polymerization seems to have little, if any, effect on the chain microstructure. This was shown[17] for the case of isoprene over a range of temperature from -25 to 46°C for the undiluted monomer, and from -25 to 60°C for *n*-hexane solutions. A similar lack of effect was found[17] for the polymerization of undiluted butadiene in the temperature range from 0 to 35°C. It was noticed, however, that the use of temperatures higher than 50°C led to some side reactions, evidenced by the appearance of a yellowish tint to the solution. This may be the reason for the small but significant changes in microstructure reported[19] for polyisoprene prepared at elevated temperatures.

The influence of polar solvents, even in trace amounts, on the isomeric chain unit structure of the polydienes, as described above, can now be rationalized by recent information concerning the structure of the carbon–lithium bond. The more subtle effects of initiator and monomer concentration on the chain structure obtained in nonpolar media poses a more perplexing problem, for which a definitive answer has not yet been forthcoming, although some tentative hypotheses can be considered. This will be discussed in the following section.

2. The Nature of the Propagating Carbon–Lithium Bond in the Polymerization of 1,3-Dienes

In view of the nonterminating character of the propagation reaction in these polymerizations, it has actually been found possible to study the structure of the propagating chain end, i.e., the carbon–lithium bond. NMR techniques, mainly proton resonance, have been successfully used for this purpose in recent years. The dienes included in these investigations were: butadiene,[22–24] isoprene,[22,25,26] 2,3-dimethylbutadiene,[22,26] several pentadienes,[27,28] and 2,4-hexadiene.[27,28] The proton resonance spectra were obtained from solutions containing about 1 *M* concentrations of chain ends, and for maximum precision, using oligomeric species to maintain low viscosities and prevent peak broadening.

In addition to the spectra of the chain-end protons, attempts were made[22,24,26] to note any perturbation of the monomer protons in the presence of chain ends, i.e., any signs of complexation between the monomers and the propagating chain ends. No signs of such intermediate complex formation were noted, indicating that, if such a monomer–lithium complex is formed, it must have too short a lifetime for detection by the methods used. However, it is noteworthy in this regard that NMR studies on 3-butenyllithium,[29,30] which forms a hexameric aggregate, show clear evidence of perturbation of the olefinic double bond by the presence of the lithium, presumably by dipole–dipole interactions. Hence it remains to find a way to detect such apparently fugitive complexes between the 1,3-dienes and lithium in polymerization systems.

a. Butadiene, Isoprene, and 2,3-Dimethylbutadiene

The type of proton resonance spectra obtained for the chain-end units of butadiene, isoprene, and 2,3-dimethylbutadiene in benzene are shown in Figs. 7.3 and 7.4. In this work, the technique of "pseudotermination" was used instead of the usual termination by water or alcohol. In this technique, the "transparent" (to NMR) butadiene-d_6 was added to the chain end in order to remove the lithium from the carbon atoms without introducing polar molecules which might affect the bond structures. The butadiene-d_6 was also used to note the effect of chain length on chain-end structure without introducing a high concentration of in-chain protons.

A comparison of the spectra of the active versus the pseudoterminated species in Figs. 7.3 and 7.4 clearly identifies the chain-end protons, i.e., those

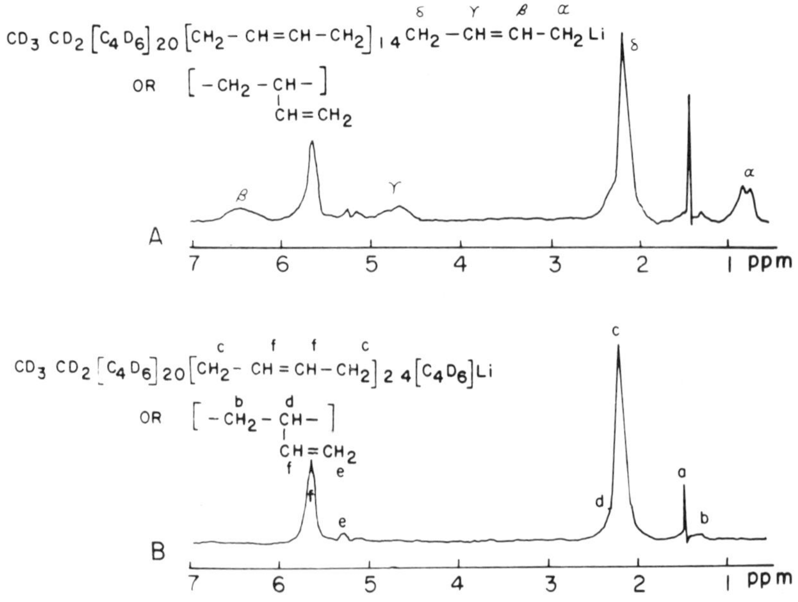

Fig. 7.3 100 MHz NMR spectra[16,22,24] of polybutadienyllithium in benzene-d_6 at 23°C. (A) Active; (B) pseudoterminated (a, cyclohexane, internal reference). (Courtesy of John Wiley and Sons.)

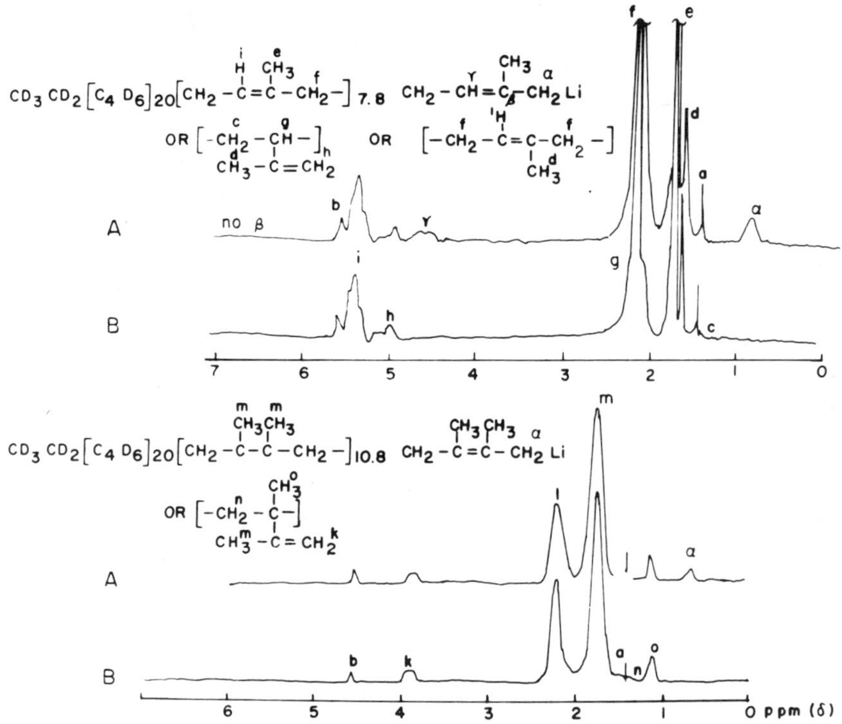

Fig. 7.4 100 MHz NMR spectra of polyisoprenyllithium (top) and poly-2,3-dimethylbutadienyllithium (bottom) in benzene-d_6 at 23°C. (A) Active; (B) Pseudoterminated.[16, 22, 26] (a, internal reference cyclohexane; b, remnant nondeuteration in perdeuterobutadiene units.) (Courtesy of John Wiley and Sons.)

belonging to the last unit of the chain, and which are attached to the four carbon atoms marked α, β, γ, and δ. It can be noted that the δ protons appear to be sufficiently far removed from the lithium atom not to undergo any perturbation, so that they correspond to their in-chain analogs (peak c). Such is not the case, obviously, for the α, β, and γ protons. The initiator used, ethyllithium-d_5, was not sufficiently reactive to be entirely consumed at the monomer–initiator ratio used, so that there was some residual initiator, which would be presumed to form a cross-complex with the chain ends (see Chapter 5). In order to avoid this complication, some later work was done with *sec*-butyllithium as the initiator for isoprene and 2,3-dimethylbutadiene, since it is known as a very fast initiator with dienes, leading to very low oligomers and even telomers, if desired. It should be noted, however, that the 20-unit polybutadienyllithium prepared from the C_2D_5Li and perdeuterobutadiene was done to avoid the unusual chain microstructures (high 1,2) and large aggregates reported[31] for the low oligomers of butadiene.

From the spectra shown in Figs. 7.3 and 7.4 it is possible to draw the following conclusions:

(i) The polybutadienyllithium chain end is apparently 100% 1,4 with no 1,2 structure observable, even though the in-chain units contain about 9% 1,2 structures

(ii) The polyisoprenyllithium chain end is exclusively 4,1 (no β protons) with no 4,3 structure detectable, even though the in-chain units have about 10% of 3,4 structures

(iii) The bond between the α-carbon and the lithium is covalent (localized) for both butadiene and isoprene, as attested to by the fact that the two α protons are equivalent (rotation around the α-carbon–β-carbon single bond), and that there is no perturbation of the γ protons, which would be expected for a delocalized (π-allyl) type bond

(iv) The polydimethylbutadienyllithium chain ends also appear to be exclusively 1,4, even though the chains contain 15–20% of 1,2 in-chain units; and the carbon–lithium bond is again σ-type rather than π-type.

The simplest way to account for the presence of vinyl-type in-chain units in the absence of any such units as chain ends would be to postulate an undetectably small amount of delocalized (π-bonded) carbon–lithium chain ends in equilibrium with the observed σ-bonded chain ends. This type of equilibrium is depicted in Fig. 7.5, where the π-bonded chain ends can presumably lead to vinyl-type in-chain units whenever an incoming monomer molecule attacks the γ-carbon rather than the α-carbon. Thus the undetectably small amount of π-bonded chain ends can result in a noticeable vinyl in-chain

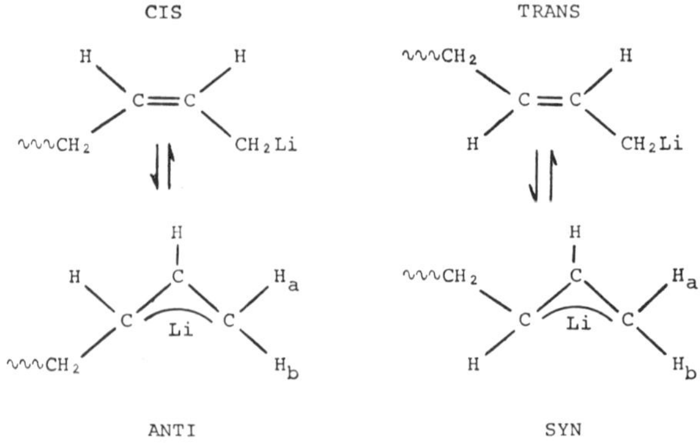

Fig. 7.5 Localized–delocalized equilibrium in polybutadienyllithium chain ends.[16,22,24] (Courtesy, Rubber Division, American Chemical Society.)

content, depending on the relative reactivity of the σ-bonded and π-bonded species.

It can be noted from Fig. 7.5 that this type of σ–π equilibrium should not lead to isomerization of the β–γ double bond (cis–trans), and this was indeed found to be the case,[22,26] at least in nonpolar media. In polar solvents, on the other hand, such isomerization was found to occur, and this will be discussed later. It should be pointed out, however, that there is some controversy about the possibility of cis–trans isomerization in nonpolar media, claims having been made that such a reaction has been observed.[25]

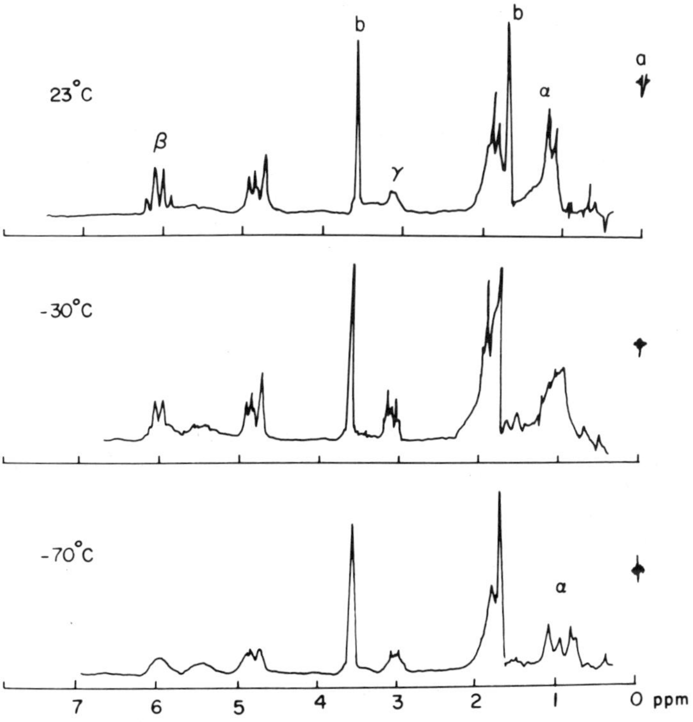

Fig. 7.6 100 MHz NMR spectra of polybutadienyllithium in H_4-furan-d_8 at various temperatures.[16,22] (a, TMS; b, nondeuteration in THF-d_8.) (Courtesy of John Wiley and Sons.)

The type of equilibrium proposed in Fig. 7.5 to account for the formation of side-vinyl in-chain units receives support from studies of chain-end structure in polar media. Figure 7.6 shows spectra of polybutadienyllithium in H_4-furan at three different temperatures. This is the kind of system which leads to a preponderance of 1,2 units in the polybutadiene (see Table 7.3), and this appears to coincide with the π character of the carbon–lithium bond. Evidence for the latter in Fig. 7.6 can be summarized as follows:

(i) Two distinguishable α protons can be observed (two overlapping doublets), indicating a restricted rotation of the α–β partial double bond

(ii) The γ protons are located upfield at 3.3 ppm instead of at 4.7 ppm as in Fig. 7.3 for benzene solutions, clearly showing the effect of the lithium atom close to the γ position.

All of the above evidence for the existence of σ- and π-bonded lithium and its relation to the ratio of 1,4 to side-vinyl chain units in the polymers makes it possible to suggest a mechanism for the effect of solvents on the chain structure. This is shown in Fig. 7.7 for butadiene, where 1,4-addition results from a concerted four-center reaction between the monomer and the σ-carbon–lithium bond, and 1,2-addition occurs when the monomer reacts with the γ-carbon of the π bond. It is also proposed that a 1,2 σ-bonded chain end actually exists as one of the equilibrium forms in the presence of polar solvents (ethers, etc.), since this can account for the isomerization of the β–γ bond (cis–trans) which was found to occur in these cases.

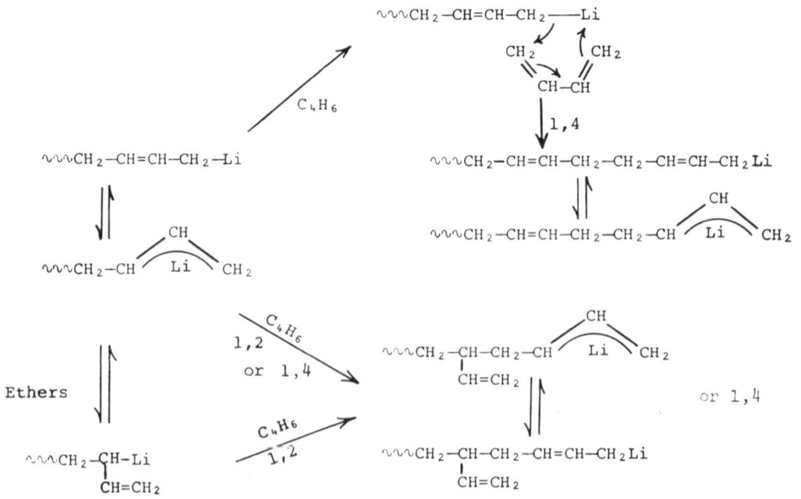

Fig. 7.7 Proposed mechanism for chain propagation in organolithium polymerization of butadiene.[16] (Courtesy, Rubber Division, American Chemical Society.)

It is interesting to note that additional support for the general features of this proposed mechanism comes from a very recent study[32] of the ^{13}C spectra of polybutadienyllithium in hydrocarbon media.

b. Terminally Substituted 1,3-Dienes

NMR studies of pentadienes and hexadiene have provided additional confirmation[28] of the mechanism described in Fig. 7.7. Thus Fig. 7.8 shows the proton resonance spectrum of oligo-1,3-pentadienyllithium prepared by direct initiation with ethyllithium-d_5. The α, β, and γ peaks appear in the same locations as for butadiene and isoprene in nonpolar solvents and again suggest only a 4,1 localized chain end as in isoprene. The in-chain units, on the other hand, consist of 80–85% 1,4 and 15–20% 1,2 units with no evidence of 3,4 structures. Since 1,2 in-chain units cannot be formed from a 4,1 chain end, it can only be assumed that they arise from the undetectable presence of some 1,4 chain ends having a delocalized structure as in the case of butadiene or isoprene. Actually, direct evidence was found for the existence of such chain ends, as will be seen later.

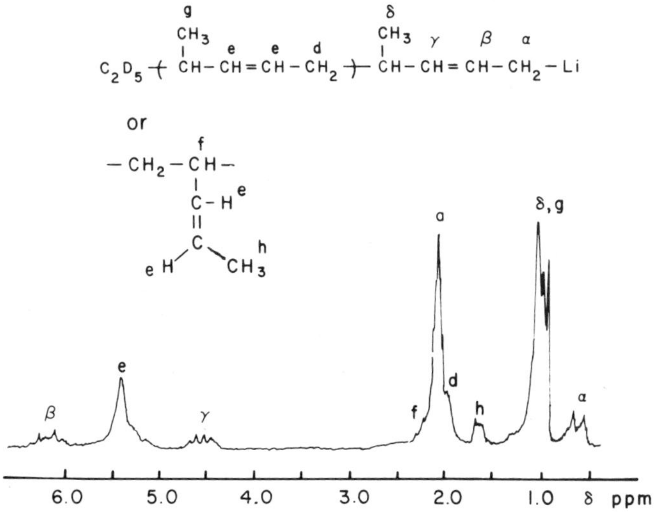

Fig. 7.8 100 MHz NMR spectrum of poly-1-*trans*-3-pentadienyllithium in toluene-d_8 at 23°C.[16,28] (a, nondeuteration in toluene-d_8.) (Reprinted with permission from Ref. 28. Copyright 1973 American Chemical Society.)

However, it is the structure of the 1,2 in-chain units which supplies convincing evidence to corroborate the mechanism described in Fig. 7.7. Thus, although the starting monomer was 1-*trans*-3-pentadiene, the 3,4 double bonds in the 1,2 in-chain units had 25–35% cis structure. Direct isomerization of the monomer could not provide the explanation, since it was not observed. Instead, this change can easily be explained by isomerization which can occur as a result of the σ–π bond equilibrium of the type described in Fig. 7.7. This type of bond equilibrium for a 1,4 chain end of pentadiene is shown in Fig. 7.9

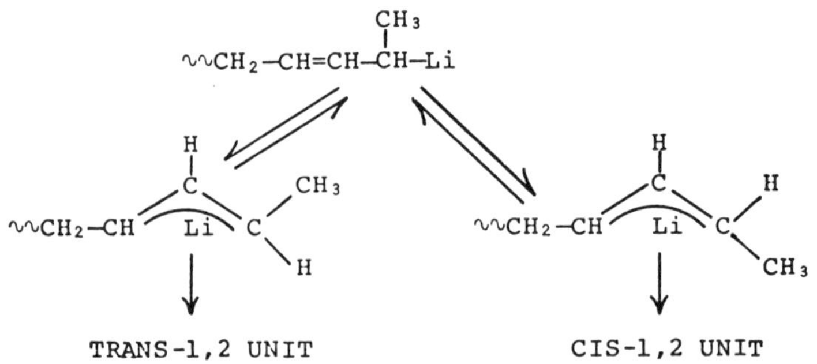

Fig. 7.9 Mechanism of the isomerization of the 3,4 bond in 1-*trans*-3-polypentadienyllithium during polymerization.[16,28] (Courtesy, Rubber Division, American Chemical Society.)

and clearly shows how the 3,4 double bond can be isomerized. In other words, whereas the σ–π equilibrium cannot lead to an isomerization of the 1,4 structure, it can do so for the 3,4 double bonds in a pentadiene.

The scheme proposed in Fig. 7.9 is based on the presence of 1,4 chain ends, which are in turn postulated because of the finding of 15–20% 1,2 in-chain units in the polymer. Fortunately, in this case it is actually possible to find direct evidence for 1,2 chain-ends. This is found in Fig. 7.10, which shows proton resonance spectra of an oligo-1-*trans*-3-pentadiene prepared with the more

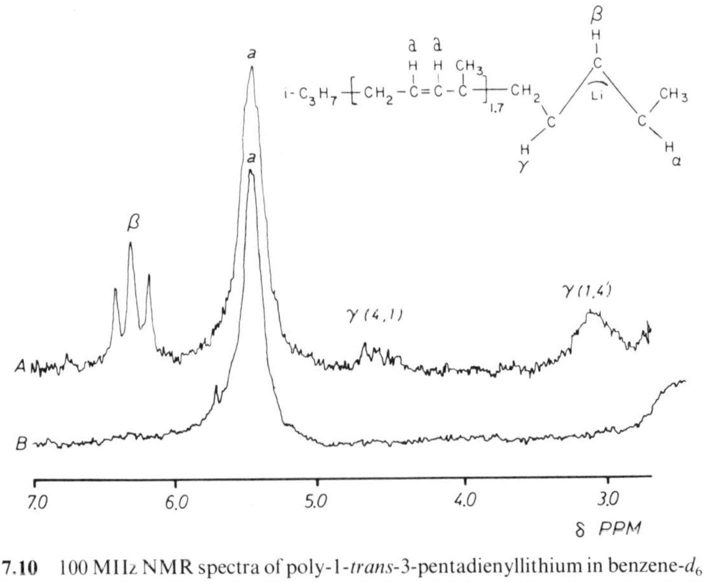

Fig. 7.10 100 MHz NMR spectra of poly-1-*trans*-3-pentadienyllithium in benzene-d_6 at 23°C. (A) Active; (B) pseudoterminated.[16] (Courtesy, Rubber Division, American Chemical Society.)

reactive initiator isopropyllithium-d_7, leading to an even shorter chain length than with the ethyllithium. These lower oligomers actually show the presence of 1,4 chain ends (at 3.1 ppm) as well as the 4,1 chain ends at 4.6 ppm, as before. Apparently the initiation step can involve attack either at the 1 or 4 position of the diene, leading to *secondary* or *primary* carbanions, respectively. However, further propagation apparently favors attack at the 4 position, presumably because this leads to formation of the more stable *primary* carbanions.

It is interesting to ncte that the γ protons of the 1,4 chain ends visible in Fig. 7.10 occur upfield at 3.1 ppm, corresponding quite closely to the position of the γ protons of the delocalized chain end of polybutadienyllithium in H_4-furan (Fig. 7.6). Hence these 1,4 chain ends must be, at least in part, in a delocalized state, even though they are in a nonpolar medium. Apparently the relative instability of the secondary carbanion at the α-carbon in a 1,4 chain end provides the driving force toward stabilization by resonance. This also corroborates the mechanism proposed in Fig. 7.7, since it shows that delocalization can occur in a nonpolar solvent. Incidentally, it should be noted that the 4,1 chain ends clearly shown both in Fig. 7.8 and 7.10 apparently do not lead to any 3,4 in-chain units, even though some delocalization can be assumed to occur. It is possible that the presence of a bulky methyl group on the δ-carbon inhibits addition of the monomer at the γ carbon.

As stated above, the 1,4 chain ends in the pentadiene can only be detected at the initiation stages, changing to the more thermodynamically stable 4,1 chain ends as the chains grow. No such stabilization is possible in the case of 2,4-hexadiene, where only one type of 1,4-addition is possible, leading to a

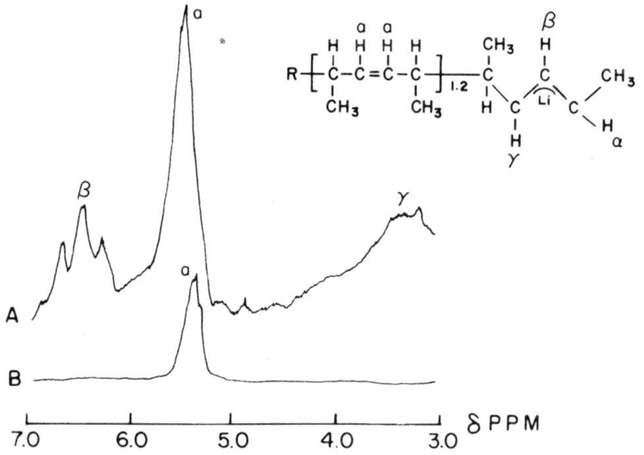

Fig. 7.11 60 MHz NMR spectra of the olefinic protons of poly-2,4-hexadienyllithium in benzene-d_6 at 23°C. (A) Active; (B) pseudoterminated.[16,28] (Reprinted with permission from Ref. 28. Copyright 1973 American Chemical Society.)

substituted or secondary carbanion. Since this type of chain end was seen only in the delocalized form in Fig. 7.10, it is not surprising that it appears in the same way in Fig. 7.11 for the 2,4-hexadiene. It can be noted that no detectable amount of the covalent form of the γ protons can be seen at 4.7 ppm in this case. Instead, there is clear evidence for the existence of a delocalized carbon–lithium bond in a nonpolar solvent, and this reinforces the general mechanism proposed in Fig. 7.7. It should also be mentioned that, despite the high electron density at the γ-carbon, no 2,3 in-chain structures were detectable, only 2,5 chain units being observed. This is the same phenomenon noted in the case of the pentadiene, and confirms the idea that a methyl substituent at the δ position apparently blocks any monomer addition at the γ carbon. This is an interesting observation, since a methyl group at the β position, as in isoprene, apparently does not suppress the activity of the γ carbon in the same way.

The existence of these secondary carbanions as delocalized species in nonpolar media raises some interesting questions about association of chain ends. It has been shown previously that organolithium molecules show various states of association in hydrocarbon solvents, but not in polar solvents such as ethers. Thus primary alkyllithiums are more highly associated than secondary alkyllithiums (see Table 3.2), while the primary carbanions obtained from butadiene and isoprene are associated in pairs in hydrocarbon solvents but not in ethers such as H_4-furan (see Chapter 5). Hence the question arises as to the association of the above oligohexadienyllithium, which apparently has a delocalized carbon–lithium bond and even exhibits the red-colored solutions usually found in ether solvents. Actually, the association of these species was measured[27] by the viscometric method described in Chapter 5 and, as expected, the delocalization of the carbon–lithium bond had a strong effect on this phenomenon. Thus the chain ends were found to be much less associated than in butadiene or isoprene, the dimer content being only about 30%, the remaining 70% of the chain ends being in the unassociated state.

3. Chain Structure and Mechanism of Polymerization

Our present knowledge about the chain structure of dienes and the nature of the propagating chain ends in these polymerizations, as described above, can help to elucidate the mechanism, at least in part. Thus the mechanism proposed in Fig. 7.7 can account for the well-known effect of polar solvents on the side-vinyl content of the chains. It can also help to rationalize the effect of solvents on the copolymerization of these dienes with styrene, as described in Chapter 6. Thus, in hydrocarbon solvents it would obviously be more difficult for a styrene molecule to participate in the four-center reaction with the predominant σ-bonded polydienyllithium chain ends. Hence the inclusion of the diene monomer would be favored, as observed. However, in the presence of polar solvents, the carbon–lithium bond becomes delocalized and a four-center reaction is no longer predominant, so that the styrene can react vigorously with the carbanionic chain ends. Thus the increase in the inclusion of styrene parallels the increase in side-vinyl content of the chains, both being the result of bond delocalization at the chain end. It should be noted, too, that in hydrocarbon media the chain ends are associated as dimers, which can perhaps introduce additional restrictions for monomer entry, while this is not the case in polar solvents.

There are two additional phenomena in the organolithium polymerization of dienes in hydrocarbon solvents which the above proposed mechanism cannot rationalize, i.e., the very high cis-1,4 content of polyisoprene compared to polybutadiene and the low kinetic orders of the propagation rate (see Chapter 4). Since it is known that the 4,1 polyisoprene chain ends have a high proportion of cis content while the reverse is true for the 1,4-polybutadiene chain ends, analogous to their in-chain structure, it would be tempting to conclude that the configuration of these chain units depends on the conformation of the monomer molecules at the time of monomer addition. However, the best recent evidence from microwave spectroscopy,[33] indicates that both of these monomers have a predominant transoid conformation. Hence the propagation step itself must control the way in which the monomer enters.

In this connection, a recently proposed theory[34] to account for the effect of lithium concentration on chain structure (see Tables 7.4 and 7.5, and Figs. 7.1 and 7.2) deserves mention. This proposed mechanism is based on a competition between the rates of isomerization and propagation of the growing chain ends of polyisoprene and polybutadiene. Thus each monomer addition presumably leads to a cis chain end which then isomerizes at a certain rate to the trans form. Hence the final in-chain structure of the given chain-end unit depends on the length of the interval between monomer additions. As the concentration of initiator is increased, i.e., the concentration of chain ends increases, this interval becomes *longer*, due to the very low kinetic order of the propagation step, i.e., $\frac{1}{4}$ or lower. In other words, at higher initiator concentration the chain ends must wait for a longer period before reacting with monomer, and hence have a greater opportunity to isomerize to the trans form. This can then account for the higher trans content at higher initiator concentration, found for both isoprene and butadiene.

This interesting hypothesis is, of course, very attractive, since it can account for both the well-established effect of lithium concentration as well as monomer concentration on the chain structure (higher monomer concentration would lead to faster propagation and hence higher cis content). However, it leaves unresolved the following questions: (i) evidence of cis–trans isomerization in nonpolar solvents was found by one group of workers[25] in contradiction to the findings of another group[26]; (ii) the experimental studies on the cis–trans isomerization were carried out[34] on model compounds (e.g., 2,2,5-trimethyl-2-hexenyllithium) at relatively high concentrations (0.04 – 0.3 M) where these organolithium compounds would probably be associated as tetramers,[23] rather than in the dimeric state which occurs at the lower concentration used in polymerization, and this could have a great effect on isomerization possibilities; and (iii) no change in chain structure was found with degree of conversion of monomer to polymer[17] up to very high conversions (> 80%) where the propagation rate would be greatly reduced and should presumably lead to a higher degree of isomerization.

An alternative mechanism for the effect or organolithium concentration on chain structure might very well be related to the kinetics of propagation; i.e., the very low orders found, varying from $\frac{1}{4}$ to $\frac{1}{6}$ and depending on the particular hydrocarbon solvent (see Figs. 4.5 and 4.6). Thus, as stated previously, these fractional orders have been erroneously ascribed to an association equilibrium in which the chain ends are associated as tetramers or hexamers, as the case may be, and are unable to propagate, only the small equilibrium amount of free, unassociated chain ends being active. Unfortunately for this interpretation, experimental evidence points unequivocally to a dimeric state of association of the chain ends in all of these cases (see Chapter 5). Furthermore, a more acceptable hypothesis would invoke a *direct* first-order reaction between the monomer and the *associated* chain ends, the effect of chain-end concentration (i.e., the low kinetic order) simply reflecting a *change* in the structure and reactivity of the carbon–lithium bond. This type of hypothesis is reinforced by the finding[35] that, because of the highly polar nature of organolithium compounds, the dielectric constant of hydrocarbon solvents increases significantly with addition of very low concentrations of alkyllithium. Unfortunately, it has not yet been found possible to apply NMR techniques to a study of these carbon–lithium bond structures at these very low concentrations ($\sim 10^{-2}$–10^{-5} M).

References

1. F. E. Matthews and E. H. Strange, British Patent 24,790 (1910).
2. C. D. Harries, *Justus Liebigs Ann. Chem.* **383**, 157 (1911).
3. C. D. Harries, U. S. Patent 1,058,056 (1913).
4. A. W. Meyer, R. R. Hampton, and J. A. Davison, *J. Am. Chem. Soc.* **74**, 2294 (1952).
5. K. Ziegler, F. Dersch, and H. Willthan, *Justus Liebigs Ann. Chem.* **511**, 13 (1934).
6. F. W. Stavely and co-workers, *Ind. Eng. Chem.* **48**, 778 (1956).
7. F. C. Foster and J. L. Binder, *Adv. Chem. Ser.* **17**, 7 (1957).
8. H. Morita and A. V. Tobolsky, *J. Am. Chem. Soc.* **79**, 5853 (1957).
9. A. V. Tobolsky and C. E. Rogers, *J. Polym. Sci.* **40**, 73 (1959).
10. R. S. Stearns and L. E. Forman, *J. Polym. Sci.* **41**, 381 (1959).
11. M. Morton and L. J. Fetters, *J. Polym. Sci., Part A* **2**, 3311 (1964).

12. C. A. Uraneck, *J. Polym. Sci., Polym. Chem. Ed.* **9**, 2273 (1971).
13. T. A. Antkowiak, A. E. Oberster, A. F. Halasa, and D. P. Tate, *J. Polym. Sci., Polym. Chem. Ed.* **10**, 1319 (1972).
14. S. Bywater, *Adv. Polym. Sci.* **4**, 66 (1965).
15. L. E. Forman, *in* "Polymer Chemistry of Synthetic Elastomers" (J. P. Kennedy and E. Tornqvist, eds.), Part II, pp. 522, 567–569. Wiley (Interscience), New York, 1969.
16. M. Morton and L. J. Fetters, *Rubber Chem. Technol.* **48**, 359 (1975).
17. J. Rupert, Ph.D. Dissertation, University of Akron, Akron, Ohio, 1975; see also M. Morton and J. P. Rupert, ACS Symposium Series, American Chemical Society, Washington, D.C. (in press).
18. F. Schue, D. J. Worsfold, and S. Bywater, *Macromolecules* **3**, 509 (1970).
19. D. J. Worsfold and S. Bywater, *Can. J. Chem.* **42**, 2884 (1964).
20. E. R. Santee, Jr., R. Chang, and M. Morton, *J. Polym. Sci., Polym. Lett. Ed.* **11**, 449 (1973).
21. E. R. Santee, Jr., L. O. Malotky, and M. Morton, *Rubber Chem. Technol.* **46**, 1156 (1973).
22. M. Morton, R. D. Sanderson, and R. Sakata, *J. Polym. Sci., Part B* **9**, 61 (1971).
23. W. H. Glaze, J. E. Hanicak, M. L. Moore, and J. Chadhuri, *J. Organomet. Chem.* **44**, 39 (1972).
24. M. Morton, R. D. Sanderson, and R. Sakata, *Macromolecules* **6**, 181 (1973).
25. F. Schue, D. J. Worsfold, and S. Bywater, *J. Polym. Sci., Part B* **7**, 821 (1969); *Macromolecules* **3**, 509 (1970).
26. M. Morton, R. D. Sanderson, R. Sakata, and L. A. Falvo, *Macromolecules* **6**, 186 (1973).
27. M. Morton, L. A. Falvo, and L. J. Fetters, *J. Polym. Sci., Part B* **10**, 561 (1972).
28. M. Morton and L. A. Falvo, *Macromolecules* **6**, 190 (1973).
29. J. P. Oliver, J. B. Smart, and M. T. Emerson, *J. Am. Chem. Soc.* **88**, 401 (1966).
30. J. B. Smart, R. Hogan, P. A. Scherr, M. T. Emerson, and J. P. Oliver, *J. Organomet. Chem.* **64**, 1 (1974).
31. H. S. Makowski and M. Lynn, *J. Macromol. Chem.* **1**, 443 (1966).
32. A. F. Halasa, D. N. Schulz, D. P. Tate, and V. D. Mochel, *Adv. Organomet. Chem.* **18**, 55 (1980).
33. S. L. Hsu, M. K. Kemp, J. M. Pochan, R. O. Benson, and W. H. Flygare, *J. Chem. Phys.* **50**, 1482 (1969).
34. D. J. Worsfold and S. Bywater, *Macromolecules* **11**, 582 (1978).
35. M. Morton, unpublished results.

Part II
Applications of Anionic Polymerization in Polymer Synthesis

Chapter

8 Molecular Weights and Their Distribution

1. Anionic Polymerization and the Poisson Distribution of Molecular Weights

Although Dostal and Mark,[1] as early as 1935, had considered the effect of the rates of initiation and propagation on the molecular weight distribution in a nonterminating addition polymerization, it was undoubtedly Flory[2] who first treated such a system quantitatively. Basing his treatment on the anionic polymerization of ethylene oxide by sodium alkoxides, where the rate for the initiation step is of the same order of magnitude as that of the propagation step and there is no termination or transfer, he derived the relations which predicted a very narrow (Poisson) distribution for such polymers at high molecular weights. In his derivation the molecular weight distribution in such a system is defined by the expressions

$$P_j = e^{-x}x^{j-1}/(j-1)! \tag{8.1}$$

$$W_j = [x/(x+1)]je^{-x}x^{j-2}/(j-1)! \tag{8.2}$$

where P_j and W_j are the number and weight fraction of j-mers and x denotes the number of monomer units which have reacted per initiator molecule. Hence H, the ratio of the weight average (x_w) to number average (x_n) chain length, is given by

$$H = x_w/x_n = 1 + (x_n - 1)/x_n^2$$
$$= 1 + 1/x_n \quad \text{at high values of } x \tag{8.3}$$

Equation (8.3) predicts that, when x is 100, the ratio H is equal to 1.01, so that the polymer is virtually "monodisperse." It should be remembered, of course, that this prediction is based on the assumptions that: (i) there is no termination or transfer during chain growth, (ii) the initiation step is at least as rapid as the propagation step, and (iii) all the monomer molecules have equal probability of reacting (perfect mixing). In practice, even when conditions (i) and (ii) are possible, the requirement of perfect mixing is difficult to meet. Furthermore, there is no available method for the measurement of x_w/x_n with sufficient accuracy to confirm a value as low as 1.01, should it actually be attained, the lowest values currently measured being about 1.03 to 1.05. This is due to the fact that direct methods of measuring x_w and x_n, i.e., light scattering and osmometry, respectively, can at best be done with an accuracy of about 2%, so that their ratio has an even higher error.

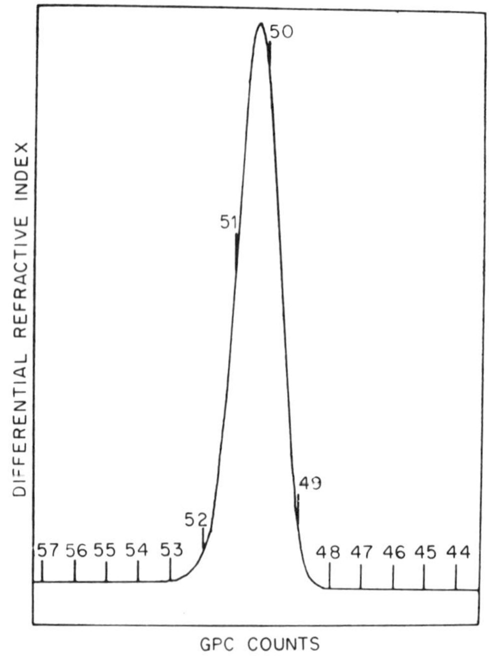

Fig. 8.1 Gel permeation chromatogram of polystyrene prepared with *sec*-butyllithium in cyclohexane at 35°C.[3] $M_s = 9.6 \times 10^4$; $M_n = 9.7 \times 10^4$; $M_w = 9.8 \times 10^4$. (Courtesy of Rubber Division, American Chemical Society.)

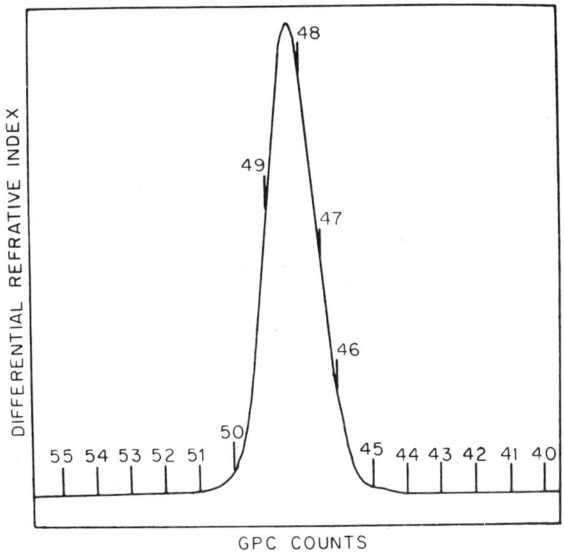

Fig. 8.2 Gel permeation chromatogram of polyisoprene prepared with *sec*-butyllithium in cyclohexane at 35°C.[3] $M_s = 1.3 \times 10^5$; $M_n = 1.35 \times 10^5$; $M_w = 1.4 \times 10^5$. (Courtesy of Rubber Division, American Chemical Society.)

Gel permeation chromatography, when properly done,[3] probably provides the most accurate method for obtaining x_w/x_n of a polymer, although "column broadening" could actually tend to give slightly high values. Even so, some very narrow molecular weight distributions have been attained in homogeneous anionic polymerization, as shown by gel permeation chromatography. Figures 8.1 and 8.2 show examples of gel permeation chromatograms of a polystyrene and a polyisoprene prepared in this way to an extremely narrow molecular weight distribution. These figures also demonstrate another feature of such polymerizations, i.e., the predictable values of the stoichiometric molecular weight M_s based on the ratio of monomer polymerized per initiator molecule.

2. Factors Affecting the Molecular Weight and Its Distribution

a. Effect of Rates of Initiation and Propagation

It has already been pointed out that, in order to obtain a narrow molecular weight distribution, it is necessary to avoid any chain termination (or transfer) processes and to have a relatively rapid initiation rate compared to propagation. These factors have been considered and discussed in some detail in the literature. Thus, in some early work Magat[4] developed some quantitative relations between the molecular weight distribution and the ratio of the propagation to initiation velocities in a termination-free polymerization. Some time later, Gold[5] also calculated the effect of the propagation/initiation rate ratios on the M_w/M_n ratio and concluded that, even in an extreme case (e.g., $k_p/k_i = 10^6$), the latter ratio should not exceed 1.3–1.4. Bresler *et al.*[6] discussed the anionic polymerization of butadiene and isoprene with regard to the effect of propagation and initiation rates on the MWD. Coleman and Fox[7] treated the special case of a nonterminating polymerization in which the growing chain end has two forms in dynamic equilibrium, each with its own propagation rate. This type of situation was also considered by Figini[8] and by Szwarc and Hermans.[9] Eisenberg and McQuarrie,[10] on the other hand, showed how any desired polydispersity in molecular weights could be obtained by controlled addition of initiator during the polymerization.

b. Effect of Chain Termination or Transfer

In considering the effect of "chain termination" processes in anionic polymerization, a distinction must be made between the following possibilities:

(i) partial destruction of initiator by impurities;
(ii) termination of growing chains by impurities, monomer, solvent, or bond rearrangements; and
(iii) termination of growing chains by a transfer reaction.

The process described in (i) above should not, by itself, affect the molecular weight distribution but only the predicted molecular weight based on monomer/initiator stoichiometry, the actual molecular weight always being higher than predicted. However, this only holds for a monofunctional polymerization (e.g., organolithium), since a difunctional system (e.g., sodium naphthalene) could produce a *bimodal* distribution resulting from the destruction of only one of the initiator functional groups in some fraction of the initiator.

Process (ii) should not, in principle, affect the number-average molecular weight but will, of course, broaden the distribution, depending on the relative rates (and extents) of termination versus propagation. The effect of such termination processes has been treated theoretically by Figini,[8a] Orofino, and Wenger,[11] and Coleman, Gornick, and Weiss.[12]

As for chain transfer, which has been found to occur in certain anionic polymerization, it has received far less attention than in free radical systems. For example, there has been no compilation of transfer constants as in free radical polymerization, despite the fact that it was observed in some of the earliest studies of anionic polymerization. Thus Higginson and Wooding[13] noted that the solvent liquid ammonia was an active transfer agent in the polymerization of styrene by potassium amide, while Robertson and Marion[14] made the same observation for the action of toluene in the polymerization of butadiene by metallic sodium. It should be noted that these instances of "chain transfer" in anionic polymerization by organoalkali initiators actually represent "metallation" reactions, whereby a labile proton (in the solvent or monomer) is displaced by the metal.

A number of cases of anionic chain transfer have been reported in the literature, some of the examples representing important processes for the control of the molecular weight. Thus Bower and McCormick[15] and also Brooks[16] showed that transfer to toluene occurred when the latter was used as a solvent in the organosodium polymerization of styrene, since lower molecular weights than predicted were obtained. Similarly, lower molecular weights were found[17,18] in the polymerization of ethylene by *n*-butyllithium in benzene in the presence of tetramethylethylenediamine (TMEDA), because the latter apparently promoted a chain transfer reaction with the benzene, as evidenced by the presence of phenyl groups in the polymer. When the solvent was *n*-hexane instead of benzene, there was no evidence for any chain transfer,[19] although only low molecular weights were obtained. It should be noted in this connection that spontaneous termination can occur under certain conditions in organoalkali polymerizations by the decomposition of the growing chain end into an alkali hydride (e.g., LiH or NaH) and a terminal double bond, the hydride apparently being incapable of initiating new chains.

The occurrence of chain transfer in the polymerization of butadiene by organoalkali initiators, first noted with toluene as solvent in sodium polymerization,[14] has been applied to other systems involving this important monomer. Thus organosodium polymerization of butadiene in a toluene–H_4-furan mixture was found to exhibit so much chain transfer that only oligomers of 10 units or less were obtained.[20] Organolithium polymerization of butadiene in the presence of toluene has been shown to produce low molecular weight polymers of any desired molecular weight or chain microstructure. Diamines and potassium *tert*-butoxide are used to promote the transfer reaction and to control the microstructure.[21–24] Copolymers of butadiene and α-methylstyrene have also been prepared in this way. It should be noted, in this connection, that such chain transfer processes, being random in nature, not only drastically reduce the *average* molecular weight, but also broaden the distribution, which tends to approach the "most probable" type, i.e., where $x_w/x_n = 2$.

The occurrence of transfer to *monomer* has been postulated in the case of the *n*-butyllithium polymerization of 9-vinylanthracene in H_4-furan, only oligomers having 4–12 units being obtained even at 100% conversion.[25,26]

Gatzke[27] showed the presence of chain transfer between polystyryllithium and toluene at 60°C and derived a relation between the molecular weight and the transfer constant, as follows:

$$x_n = [M]X/[SLi] - C_{RH}[RH]\ln(1 - X) \tag{8.4}$$

where [M] is the monomer concentration (initial), X the degree of conversion, [SLi] the chain end concentration (initiator concentration), [RH] the toluene concentration, and C_{RH} the transfer constant ($= k_{tr}/k_p$).

Equation (8.4) is presumably applicable to any case of chain transfer in a nonterminating polymerization. In the above case, the value of C_{RH} for toluene was found to be 5×10^{-6}, indicating a small extent of transfer.

3. Synthesis of Polymers with a Narrow Molecular Weight Distribution

Table 8.1 contains a listing of polymers prepared by anionic polymerization which have been reported as having near monodisperse molecular weights. The list includes those prepared by monofunctional as well as difunctional initiators, and the degree of monodispersity is indicated by the M_w/M_n ratio whenever available. For details about the synthesis and characterization of these polymers, the reader is referred to the original publications.

TABLE 8.1 Anionic Polymers with Narrow Molecular Weight Distributions

Initiator	Polymerization solvent	M_w/M_n	References
	Polystyrene		
Sodium naphthalene	H_4-furan	1.22–1.62	28
Sodium naphthalene	H_4-furan	1.06–1.12	29
Sodium naphthalene	H_4-furan	1.1	30
Sodium naphthalene	H_4-furan	1.04–1.19	31
Sodium naphthalene	H_4-furan	1.61	32
Sodium naphthalene	H_4-furan	1.04	33
Sodium naphthalene	H_4-furan	1.06–1.09	34
Sodium naphthalene	H_4-furan	1.07–1.56	35
Sodium naphthalene	H_4-furan	1.04–1.39	36
Tetra(α-methylstyrene) disodium	H_4-furan	1.05–1.09	37
α-Phenylethylpotassium	H_4-furan	1.05	38
Cumylpotassium	H_4-furan	1.1	39
Sodium biphenyl	H_4-furan	1.07	40
Sodium biphenyl	H_4-furan	1.09	41
C_2H_5Li or n-C_4H_9Li	Benzene	1.05–1.12	42
n-C_4H_9Li	Benzene[a]	1.03–1.09	43, 44

TABLE 8.1 (*continued*)

Initiator	Polymerization solvent	M_w/M_n	References
n-C_4H_9Li	H_4-furan	1.09–1.41	45
n-C_4H_9Li	Benzene/H_4-furan	1.1	46
sec-C_4H_9Li	H_4-furan	1.12, 1.17[b]	47
	Poly(vinyltoluene)		
Sodium naphthalene	H_4-furan	1.25	48
	Poly(α-methylstyrene)		
Sodium naphthalene	H_4-furan	1.0–1.03	49, 50
Tetra(α-methylstyrene) disodium	H_4-furan	1.05–1.06	51
n-C_4H_9Li	H_4-furan	1.05	52
n-C_4H_9Li + additives[c]	H_4-furan	1.00–1.02	53
	Poly(*p*-*tert*-butylstyrene)		
sec-C_4H_9Li	Benzene	1.04	54
	Poly(methyl methacrylate)		
Fluorenyllithium	Toluene	~1.0	55
Sodium biphenyl	H_4-furan	1.03–1.08	56
Lithium biphenyl	Dimethoxyethane	1.1	41
Cumylcesium	H_4-furan	<1.1	57, 58
	Polybutadiene		
n-C_4H_9Li	*n*-Hexane	1.1	59
n-, *sec*- and *tert*-C_4H_9Li	Cyclohexane	1.10–1.20	60
n-C_4H_9	Hexane	1.09–1.47	61, 62
	Polyisoprene		
n-C_4H_9Li	*n*-Hexane	1.05–1.18	63
n-C_4H_9Li	*n*-Heptane	1.1	64
n-C_4H_9Li	Cyclohexane	1.10	65
sec-C_4H_9Li	*n*-Hexane	1.1	66
n-, *sec*-, and *tert*-C_4H_9	Cyclohexane	1.13–1.35	60
sec-C_4H_9Li	Cyclohexane	1.05	67
	Polyacrylonitrile		
Na triethyl-triisopropoxyaluminate	Dimethylformamide	1.1	68
	Poly(*tert*-butyl crotonate)		
2-Methylbutyllithium	H_4-furan	1.01	69
	Poly(1-vinylpyrene)		
Cumylpotassium	H_4-furan	1.14–1.30	70
	Poly(*n*-butyl isocyanate)		
n-C_4H_9Li	Toluene	1.1	71, 72
Fluorenylsodium	Toluene	1.1	71, 72
	Poly(propylene sulfide)		
Na naphthalene	H_4-furan	<1.1	73–75
n-C_4H_9Li	H_4-furan	<1.1	76

TABLE 8.1 (*continued*)

Initiator	Polymerization solvent	M_w/M_n	References
	Poly-β-propiolactone (α-methyl and α-n-propyl)		
Tetrahexylammonium benzoate	H_4-furan	~1.1	77
	Polydimethylsiloxane		
n-C_4H_9Li or LiOH	H_4-furan, o-xylene	<1.1	78
sec-C_4H_9Li	H_4-furan, o-xylene	1.02–1.27[d]	79
Lithioacetal	Benzene/H_4-furan	~1.15	80

[a] A small amount of H_4-furan was added to accelerate initiation.
[b] M_z/M_w values for polystyrene of $M_w = 43.7 \times 10^6$ and 27.3×10^6, respectively.
[c] LiBr, LiOH, $LiOC_4H_9$.
[d] Polydispersity *increases* with mol wt.

References

1. H. Dostal and H. Mark, *Z. Phys. Chem., Abt. B* **29**, 299 (1935); *Trans. Faraday Soc.* **32**, 54 (1936).
2. P. J. Flory, *J. Am. Chem. Soc.* **62**, 1561 (1940).
3. M. Morton and L. J. Fetters, *Rubber Chem. Technol.* **48**, 405 (1975).
4. M. Magat, *J. Chim. Phys.* **47**, 841 (1950).
5. L. Gold, *J. Chem. Phys.* **28**, 91 (1958).
6. S. E. Bresler, A. A. Korotkov, M. I. Mosevitskii, and I. Ya. Poddubnyi, *Rubber Chem. Technol.* **23**, 669 (1960).
7. B. D. Coleman and T. G. Fox, *J. Am. Chem. Soc.* **85**, 1241 (1963).
8. R. V. Figini, *Makromol. Chem.* **7**, 193 (1964).
8a. R. V. Figini, *Makromol. Chem.* **44**, 497 (1961).
9. M. Szwarc and J. J. Hermans, *J. Polym. Sci., Part B* **2**, 815 (1964).
10. A. Eisenberg and D. A. McQuarrie, *J. Polym. Sci., Polym. Chem. Ed.* **4**, 737 (1966).
11. T. A. Orofino and F. Wenger, *J. Chem. Phys.* **35**, 532 (1961).
12. B. D. Coleman, F. Gornick, and G. Weiss, *J. Chem. Phys.* **39**, 3233 (1963).
13. W. C. E. Higginson and N. S. Wooding, *J. Chem. Soc.* p. 760 (1952).
14. R. E. Robertson and L. Marion, *Can. J. Res., Sect. B* **26**, 657 (1948).
15. F. M. Bower and H. W. McCormick, *J. Polym. Sci., Part A* **1**, 1749 (1963).
16. B. W. Brooks, *Chem. Commun.* p. 68 (1967).
17. G. G. Eberhardt and W. A. Butte, *J. Org. Chem.* **29**, 2928 (1964).
18. A. W. Langer, Jr., *Trans. N. Y. Acad. Sci.* [2] **27**, 741 (1965).
19. J. N. Hay, J. F. McCabe, and J. C. Robb, *J. Chem. Soc., Faraday Trans. 1* **68**, 1227 (1972).
20. S. Kume, A. Takahashi, G. Nishikawa, M. Hatano, and S. Kambara, *Makromol. Chem.* **84**, 137, 147 (1965); **98**, 109 (1966).
21. B. J. McElory and J. H. Merkley, U. S. Patent 3,678,121 (1972).
22. R. C. Morrison and C. W. Kamienski, U. S. Patent 3,725,368 (1973).
23. C. W. Kamienski and J. F. Eastham, U. S. Patent 3,742,077 (1973).
24. C. W. Kamienski and J. H. Merkley, U. S. Patent 3,751,501 (1973).
25. A. Eisenberg and A. Rembaum, *J. Polym. Sci., Part B* **2**, 157 (1964).

26. R. H. Michel and W. P. Baker, *J. Polym. Sci., Part B* **2**, 163 (1964).
27. A. Gatzke, *J. Polym. Sci., Polym. Chem. Ed.* **7**, 2281 (1969).
28. G. Meyerhof, *Z. Elektrochem.* **61**, 1245 (1957); *Z. Phys. Chem.* (*Wiesbaden*) [N. S.] **23**, 100 (1960).
29. R. Waack, A. Rembaum, J. D. Coombes, and M. Szwarc, *J. Am. Chem. Soc.* **79**, 2026 (1957).
30. H. J. Cantow, *Makromol. Chem.* **30**, 169 (1959); G. Meyerhof and H. J. Cantow, *J. Polym. Sci.* **34**, 503 (1959).
31. H. W. McCormick, *J. Polym. Sci.* **36**, 341 (1959).
32. T. Lyssy, *Helv. Chim. Acta* **42**, 2245 (1959).
33. C. Stretch and G. Allen, *Polymer* **2**, 151 (1961).
34. F. Wenger, *Makromol. Chem.* **64**, 151 (1963).
35. M. Morton, R. Milkovich, D. McIntyre, and J. L. Bradley, *J. Polym. Sci., Part A* **1**, 443 (1963).
36. F. M. Bower and H. W. McCormick, *J. Polym. Sci., Part A* **1**, 1749 (1963).
37. J. M. G. Cowie, D. J. Worsfold, and S. Bywater, *Trans. Faraday Soc.* **57**, 705 (1961).
38. S. P. S. Yen, *Makromol. Chem.* **81**, 152 (1965).
39. J. Herz, M. Hert, and C. Straszielle, *Makromol. Chem.* **160**, 213 (1960).
40. F. Wenger, *Makromol. Chem.* **36**, 200 (1960).
41. G. M. Guzman and A. Bello, *Makromol. Chem.* **107**, 46 (1967).
42. M. Morton, A. A. Rembaum, and J. L. Hall, *J. Polym. Sci., Part A* **1**, 461 (1963).
43. T. A. Altares, Jr., D. P. Wyman, and V. R. Allen, *J. Polym. Sci., Part A* **2**, 4533 (1964).
44. L. J. Fetters and M. Morton, *in* "Macromolecular Syntheses" (W. J. Bailey, ed.), p. 77. Wiley, New York, 1972.
45. S. Onogi, T. Masuda, and K. Kitagawa, *Macromolecules* **3**, 109, 117 (1970).
46. T. Masuda, Y. Ohta, and S. Onogi, *Macromolecules* **4**, 763 (1971).
47. D. McIntyre, L. J. Fetters, and E. Slagowski, *Science* **176**, 1041 (1972); *Macromolecules* **7**, 394 (1974).
48. F. M. Bower and H. W. McCormick, *J. Polym. Sci., Part A* **1**, 1749 (1963).
49. H. W. McCormick, *J. Polym. Sci.* **25**, 488 (1957); **41**, 327 (1959).
50. A. F. Sirianni, D. J. Worsfold, and S. Bywater, *Trans. Faraday Soc.* **55**, 2124 (1959).
51. F. Wenger, *J. Am. Chem. Soc.* **82**, 4281 (1960); *Makromol. Chem.* **37**, 143 (1960).
52. Roestamjah, L. A. Wall, R. E. Florin, M. H. Aldrich, and L. J. Fetters, *J. Polym. Sci., Polym. Phys. Ed.* **13**, 1783 (1975).
53. T. Kato, K. Miyaso, I. Noda, T. Fujimoto, and M. Nagasawa, *Macromolecules* **3**, 777 (1970).
54. L. J. Fetters, E. Firer, and M. Dafauti, *Macromolecules* **10**, 1200 (1977).
55. D. L. Glusker, I. Lysoff, and E. Stiles, *J. Polym. Sci.* **49**, 315 (1961).
56. A. Roig, J. E. Figueruelo, and E. Llano, *J. Polym. Sci., Part B* **3**, 171 (1965).
57. G. Lohr and G. V. Schulz, *Makromol. Chem.* **172**, 137 (1973).
58. A. H. E. Müller, H. Höcker, and G. V. Schulz, *Macromolecules* **10**, 1086 (1977).
59. J. F. Meier, Ph.D. Dissertation, University of Akron, Akron, Ohio, 1963.
60. H. L. Hsieh and O. F. McKinney, *J. Polym. Sci., Part B* **4**, 843 (1966).
61. H. E. Adams, K. Farhat, and B. L. Johnson, *Ind. Eng. Chem. Prod. Res. Dev.* **5**, 126 (1966).
62. B. L. Johnson, H. E. Adams, F. C. Weissert, and K. Farhat, *Proc. Int. Rubber Conf., 5th, 1967* p. 29 (1968).
63. M. Morton, E. E. Bostick, and R. G. Clarke, *J. Polym. Sci., Part A* **1**, 475 (1963); M. Morton, E. E. Bostick, and R. A. Livigni, *Rubber Plast. Age* **42**, 397 (1961).
64. W. H. Beattie and C. Booth, *J. Appl. Polym. Sci.* **7**, 507 (1963).
65. N. Calderon and K. W. Scott, *J. Polym. Sci., Part A* **3**, 551 (1965).
66. N. Nemoto, M. Moriwaki, H. Odani, and M. Kurata, *Macromolecules* **4**, 215 (1971).
67. L. J. Fetters and M. Morton, *Macromolecules* **7**, 552 (1974).
68. R. Chiang, J. H. Rhodes, and A. R. Evans, *J. Polym. Sci., Polym. Chem. Ed.* **4**, 3089 (1966).
69. T. Kitano, T. Fujimoto, and M. Nagasawa, *Macromolecules* **7**, 719 (1974).

70. J. J. O'Malley, J. F. Yanus, and J. M. Pearson, *Macromolecules* **5**, 158 (1972).
71. L. J. Fetters and H. Yu, *Macromolecules* **4**, 384 (1971).
72. A. J. Bur and L. J. Fetters, *Macromolecules* **6**, 874 (1973).
73. S. Boileau, G. Champetier, and P. Sigwalt, *Makromol. Chem.* **69**, 180 (1963).
74. R. S. Nevin and E. M. Pearce, *J. Polym. Sci., Part B* **3**, 491 (1965).
75. P. Sigwalt, *Chim. Ind.* (*Milan*) **96**, 909 (1966).
76. M. Morton, R. F. Kammereck, and L. J. Fetters, *Macromolecules* **4**, 11 (1971); *Br. Polym. J.* **3**, 120 (1971).
77. J. Cornibert, R. H. Marchessault, A. E. Allegrezza, and R. W. Lenz, *Macromolecules* **6**, 676 (1973).
78. C. L. Lee, C. L. Frye, and O. K. Johannson, *Polym. Prepr., Am. Chem. Soc., Div. Polym. Chem.* **10**, 1361 (1969).
79. J. G. Zilliox, J. E. L. Roovers, and S. Bywater, *Macromolecules* **8**, 573 (1975).
80. P. M. Lefebvre, R. Jerome, and P. Teyssie, *Macromolecules* **10**, 871 (1977).

Chapter

9 Block Copolymers

The synthesis of "block" copolymers has always stimulated a great deal of interest, since these generally lead to heterophase materials, i.e., finely dispersed blends of polymers which are mutually incompatible but are chemically linked to each other. As will be seen later, such materials often have unique mechanical properties. There are, of course, two basic methods of creating such polymers: (i) by forming appropriate functional groups on the chain ends of two (or more) different polymers and then linking them chemically and (ii) by converting such terminal functional groups on a polymer into polymerization initiators and then adding a second monomer, to grow a second "block" on the original chain. Method (i) has the distinct disadvantage introduced by the thermodynamic incompatibility which prevails in most mixtures of high polymers and which makes it difficult, if not impossible, to obtain intimate mixtures. Method (ii) is naturally rather complex, since it requires some sophisticated methods to convert the usual functional groups into suitable initiators of polymerization. Once that is accomplished, however, there need not be any compatibility problems since it is always possible to obtain a homogeneous solution of the polymer in the second monomer by means of a mutual solvent if necessary.

1. Synthesis by Anionic Polymerization

As pointed out in the previous chapter, it is the nonterminating character of the growing chains in anionic polymerization which can lead to an extremely narrow molecular weight distribution. This same characteristic also provides the best method for creating block copolymers by the sequential addition of different monomers. This is undoubtedly the most elegant method available, since it can produce block copolymers in which each block has a predictable molecular weight and a very narrow molecular weight distribution. Hence it is not surprising that the development of homogeneous anionic polymerization has also resulted in a phenomenal growth in block copolymers.

Among the anionic block copolymers, the most interesting and important ones have been the triblock variety, of the ABA type, since these gave rise to the rapid development of a new class of materials, the "thermoplastic elastomers." More will be said of these later in this chapter, but it might be appropriate at this time to consider the various factors involved in the synthesis of ABA triblock copolymers.

There are basically three methods[1] which can be used in the synthesis of this type of block copolymer:

(i) three-stage process using a monofunctional initiator,
(ii) two-stage process using a difunctional initiator, and
(iii) two-stage process using a monofunctional initiator plus a coupling agent.

It is instructive to compare these three methods for their efficacy in the preparation of pure block copolymers.

a. Three-Stage Process with Monofunctional Initiators

This is the type that is used with organolithium initiators, which are soluble in many solvents. It is restricted to "reversible" block copolymerization, i.e., where both the A and B blocks can initiate each other. The presence of impurities* in the monomers would lead to the following results. The addition of the first monomer would cause some loss of initiator, so that the molecular weight of the first block would be somewhat higher than expected (and so would that of the whole block copolymer). During the second stage, any impurities in the added monomer would lead to some termination of the first A blocks, so that the final product would be contaminated by such free homopolymer. In the third or final stage, any impurities in the added monomer would of course lead to formation of diblocks (AB), which would also contaminate the final triblocks. Hence the viability of the method depends on

* For the sake of simplicity, the "impurities" are restricted to those which react *instantaneously* with the initiating species, e.g., water, alcohol, or acids. Other impurities, which might react more slowly with the growing chains would affect the *distribution* of chain sizes rather than the *number* of chains.

relative effects of the inclusion of monoblocks or diblocks on the physical behavior of the final material.

b. Two-Stage Process with a Difunctional Initiator

In this case, a difunctional initiator, e.g., sodium naphthalene, is used to generate a center block B, after which monomer A is added to grow two end blocks (of equal size). This method is useful for "unilateral" block copolymerization, i.e., where the B block can initiate an A block, but not vice versa. Here the effect of monomer impurities would be as follows. The first addition of monomer B could lead to some loss of initiator, and therefore to a higher than expected overall molecular weight. However, the second stage, i.e., addition of monomer A, could result in some termination of the B blocks, either at one or both ends. The latter, i.e., termination at both ends, would lead to formation of free B polymer. However, such an occurrence would be statistically much less probable (inversely as the square of the probability of single-end termination), so that most of the terminated B blocks would still have one active end,* leading to formation of diblocks (AB).

c. Two-Stage Process with a Coupling Agent

In this case, an AB diblock is formed in the usual way, by sequential addition of A and B monomers, followed by a difunctional linking agent capable of linking together the B blocks, to form an ABA triblock. This method, too, is useful for "unilateral" systems where the A block can generate a B block but not vice versa. Like method (ii), it has the advantage of involving only two monomer additions so that the predominant effect of impurities in the monomers would be to create only free A blocks when monomer B is added. However, because of the small proportion of linking agent needed to combine the reactive chain ends, it is generally difficult to obtain exact stoichiometry, so that some proportion of free, unlinked AB diblocks[2] usually results.

2. Anionic Block Copolymers of Two or More Monomers

The above methods have been used to make block copolymers by the anionic mechanism as applied both to unsaturated monomers and heterocyclic

* This could also occur from destruction of the difunctional initiator by impurities in monomer B.

TABLE 9.1 Block Copolymers by Anionic Polymerization of Two Monomers[a]

Monomer A	Monomer B	Initiator	Solvent	Block sequence	References
Acrylonitrile	*n*-Butyl isocyanate	Na biphenyl	Toluene (10% THF)	BAB	3
		α-Phenylethyl potassium	Toluene	AB	3
	Formaldehyde	BuLi	Toluene	AB	4, 5
		Na naphthalene	THF	BAB	6
		Na biphenyl	THF	BAB	7
		BuLi	Benzene	AB	8, 9
	Ethylene oxide	Na or Li alcoholate of poly(oxyethylene) glycol	DMF	ABA	10–12
	Vinyltrimethyl(phenyl) silane	*n*-BuLi	Heptane or cyclohexane	BA	13, 14
	Propylene oxide	$M^+O^-[CH_2CH(CH_3)O]_n^- M^+$	DMF/THF	ABA	15
	Pivalolactone	NaCN	DMF	AB	16
		Bistetrabutylammonium 4,4′-azobis-4-cyano valerate	DMSO	BAB	17
	Octamethylcyclotetrasiloxane	BuLi	THF	AB	18
		Li (or Na or K) naphthalene	THF	BAB	18
	Tetrafluoroethylene sulfide	Na biphenyl	THF	BAB	19, 20
	Thiocarbonyl fluoride	Na biphenyl	THF	BAB	19, 20
	3-Vinylpyrene	BuLi	Toluene	AB	21
Methacrylonitrile	Ethylene oxide	Na or Li alcoholate of poly(oxyethylene) glycol	DMF	ABA	12
	Ethyl acrylate	BuLi/Sulfur-containing modifier		AB	22
Butadiene	Acetaldehyde	BuLi	Toluene	AB	23
		Alkali metal–aromatic complex	Toluene/THF	BAB	23
	Formaldehyde	Na naphthalene	THF	BAB	4
		BuLi	Toluene	AB	23
		Alkali metal–aromatic complex	Toluene/THF	BAB	23
		Na biphenyl	THF	BAB	24

	Acrylonitrile	BuLi	Cyclohexane	AB	25
		1,4-Dilithio-1,1,4,4-tetraphenylbutane	Cyclohexane/TMEDA (10%)	ABA	26
Butadiene (1,4-configuration)	Butadiene (1,2 and 1,4-configuration)	*s*-BuLi	Cyclohexane, followed by cyclohexane/TMEDA	ABA	27
Butadiene	*n*-Butyl isocyanate	*n*-BuLi	Toluene	AB	3
	ε-Caprolactam	*n*-BuLi	Heptane/TMEDA	AB	28
		BuLi	Toluene/Hexane	AB	29
	ε-Caprolactone	BuLi	Cyclohexane	AB	30
		Dilithium diisoprene		BAB	31
	L-Lysine NCA	Cumylpotassium	THF	AB	32
	γ-Benzyl-L-glutamate NCA	Cumylpotassium	THF	AB	33
	Carbobenzoxy-L-lysine NCA	Cumylpotassium	THF	AB	34
	Pivalolactone	Adduct of alkyllithium and *m*-diisopropenylbenzene	Cyclohexane	BAB	35
	1,1-Diphenylethylene	Li naphthalene	THF	A-(A-co-B)A[b]	36
	Ethylene sulfide	1,4-Dilithio-1,1,4,4-tetraphenylbutane	Benzene/THF	BAB	37
		Li naphthalene	THF, THF/hexane	BAB	38
		Na naphthalene	THF, THF/hexane	BAB	38
		1,4-Dilithiobutane	*n*-Heptane	BAB	39
	Propylene sulfide	Na naphthalene	THF	BAB	40–43
		Na biphenyl	THF	BAB	40
		BuLi	THF	AB	40–43
		1,4-Dilithiobutane	*n*-Heptane	BAB	39
	Thiacyclobutane	1,4-Dilithio-1,1,4,4-tetraphenylbutane	Hexane/THF	BAB	37
	Hexafluoroacetone	Na biphenyl	THF	BAB	19

(*table continues*)

TABLE 9.1 Block Copolymers – Two Monomers (*continued*)

Monomer A	Monomer B	Initiator	Solvent	Block sequence	References
Butadiene (*continued*)	Hexamethylcyclotri-siloxane	BuLi	Cyclohexane/THF	AB	44
	Isoprene	*s*-BuLi	Cyclohexane	ABA	45
		DiLi organo compounds	Isopentane	$(AB)_{2-9}$	46
		Na biphenyl	THF or hexane/THF	ABA, BAB	41, 47
		BuLi	Hexane or hexane/THF	AB	41, 47
		Na naphthalene	THF	ABA, BABAB	47
		BuLi	Benzene	BAB	48
		s-BuLi	Isopentane	AB BA × AB BA	49
	2-Vinylpyridine	BuLi	Cyclohexane	AB	25
2-Cyanobutadiene	Styrene	Φ_2PK, Φ_2PLi	THF	AB	50
	Isoprene	Φ_2NLi, Pr_2NLi	THF	AB	50
	Butadiene	BuLi	THF	AB	50
ε-Caprolactam	Formaldehyde	K + *N*-acetylcaprolactam	Me_2SO	AB	51
	Pivalolactone	Na/*N*-acetylcaprolactam	Bulk	AB	52
	β-Propiolactone	BuLi/*N*-acetyl-ε-caprolactam	Xylene	AB	53
	Poly(ε-caprolactone)-diol	Diisocyanate/Na dispersion	Bulk	ABA	54
	Poly(alkylene ether) glycol	Diisocyanate/Na caprolactam	–	ABA	55
	Polystyrene	Bisphenol A dichloroformate	Bulk	ABA	56
	Amino-ended Nylon 6-66 copolymer	Na/terephthaloyl bicaprolactam	Bulk	ABA	57
	Polysulfone	Na–caprolactam	Chlorobenzene	ABA	58, 59
		NaH dispersion	Bulk	ABA	60
Pivalolactone	DL-α-Methyl-α-*n*-propyl-β-propiolactone	Tetrahexylammonium benzoate	THF	AB	61
		Tetrabutylammonium sebacate	THF	BAB	61

DL-Lysine NCA	L-Alanine NCA	*n*-Hexylamine	DMF	ABA	62
Ethylene oxide	Amylene oxide	NaORONa	Bulk	ABA	63, 64
	Cyclohexene oxide	NaORONa	Bulk	ABA	63, 64
	Butylene oxide	RONa	Bulk	AB	65, 66
		NaORONa	Bulk	ABA, BAB	63, 64, 67
	Styrene oxide	RONa	Bulk	AB	65, 66
		NaORONa	Bulk	ABA	63, 64
	ε-Caprolactone	Na naphthalene/PEG[c]	Benzene/THF	BAB	68, 69
		NaOH/PEG[c]	–	BAB	70
	Methyl-ε-caprolactone	Bu_2Zn	Toluene	AB	71
	Pivalolactone	$Bu_4N^+OH^-$	Tolulene	AB	72
	1,1-Dimethylethylene sulfide	K carbazyl	THF	AB	73
	Ethylene sulfide	9-Carbazylpotassium	THF	AB	73
	Propylene sulfide	9-Carbazylpotassium	THF	AB	73
	n-Butyl isocyanate	Na biphenyl	Toluene (2% THF)	BAB	3
Butylene oxide	Methyl-ε-caprolactone	Bu_2Zn	Toluene	AB	71
Propylene oxide	ε-Caprolactone	Bu_2Zn	Toluene	AB	71
	Ethylene oxide	RONa	Bulk	AB	65, 66
		Bu_2Zn	Toluene	AB	74, 75
		$NaOCH_2CH(CH_3)ONa$	–	BAB	76–80
		n-Amines	–	BAB	65, 78, 81
		Ethylenediamines	–	BA AB × BA AB	65, 77
	Formaldehyde	Bu_2Zn	Benzene	AB	7
Formaldehyde	*N*,*N*-Di-*n*-butylacrylamide	*n*-BuLi	*n*-Heptane	AB	5
		n-BuLi	Benzene	AB	8
	Poly(tetramethylene oxide)	$R_4N^+OAc^-$	Cyclohexane	ABA	81
		$R_4N^+OH^-$	Cyclohexane	ABA	82

(*table continues*)

TABLE 9.1 Block Copolymers – Two Monomers (*continued*)

Monomer A	Monomer B	Initiator	Solvent	Block sequence	References
Acetaldehyde	3-Vinylpyrene	BuLi	Toluene/decalin	AB	21
Ethylene sulfide	Propylene sulfide	$Et_2Zn–H_2O$	Benzene	AB	83, 84
		K ethane dithiolate ($KSCH_2CH_2SK$)	THF	ABA	85
		Na naphthalene	THF	BAB	86
Propylene sulfide	Isobutylene sulfide	Na carbazyl		AB	87
		K (or Na) naphthalene		BAB	87
Hexamethylcyclo-trisiloxane	2,4,6-Trimethyl-2,4,6-triphenylcyclotri-siloxane	Lithium and dilithium silanolates	THF/benzene	ABA	88
	Hexaphenylcyclotri-siloxane	Φ Φ | | LiOSiOSiOLi | | Φ Φ	Benzene/ $(Me_2N)_3PO$	BAB	89, 90
		Φ | LiOSiOLi | Φ	THF/anisole	BAB	91
		Φ | LiOSiΦ | CH_3	THF/anisole	AB	91
Octamethylcyclo-tetrasiloxane	Hexaphenylcyclotri-siloxane	$(CH_3)_3SiOK$	THF	AB	92
Isoprene	Acetaldehyde	Toluene	Toluene	AB	23
		Alkali metal–aromatic complex	Toluene/THF	BAB	23

Formaldehyde	Na naphthalene	THF	BAB	4, 7
	BuLi	Benzene	AB	5–8
	BuLi	Toluene	AB	23
	Alkali metal–aromatic complex	Toluene/THF	BAB	23
	Na biphenyl	THF	BAB	24
Caprolactone	Polyisoprenyllithium	–	AB	93
	Na α-methylstyrene	–	BAB	93
	Li	–	BAB	94
Pivalolactone	Diadduct of *m*-divinylbenzene or *m*-diisopropenylbenzene and alkyllithium, e.g., *s*-BuLi	Cyclohexane	BAB	35
		Cyclohexane/THF	BAB	95
	Li	–	BAB	94
Ethylene oxide	Phenylisopropylpotassium	THF	AB	96
	Na α-methylstyrene	THF	BAB	96
Tetrafluoroethylene oxide	Na biphenyl	THF	BAB	19, 20
Ethylene sulfide	Li naphthalene	THF	BAB	38, 97, 98
	BuLi/anisole	Cyclohexane/THF	AB	99
	1,4-Dilithio-1,1,4,4-tetraphenylbutane	Hexane/THF	BAB	37
	Na naphthalene	THF	BAB	38
Propylene sulfide	BuLi	THF or hexane/THF	AB	41, 43
	Na biphenyl	THF or hexane/THF	ABA, BAB	40, 43
	Na naphthalene	THF	BAB	40–43
Thiacyclobutane	1,1-Dilithio-1,1,4,4-tetraphenylbutane	Hexane/THF	BAB	37
n-Butyl isocyanate	Na biphenyl	Toluene (2% THF)	BAB	3
Toluene diisocyanate	BuLi	Toluene/THF	AB	100
Hexamethylcyclotrisiloxane	*s*-BuLi	Cyclohexane/diglyme	AB	101

(*table continues*)

TABLE 9.1 Block Copolymers – Two Monomers (*continued*)

Monomer A	Monomer B	Initiator	Solvent	Block sequence	References
Isoprene (*continued*)	Octamethylcyclotetra-siloxane	K metal	THF	BAB	102, 103
		K naphthalene	THF	BAB	102, 103
	Piperylene	BuLi	Heptane	AB	104
	1,1-Diphenylethylene	BuLi	THF	(A–alt–B)B	36
	3-Vinylpyrene	BuLi	Toluene	AB	21
Methyl methycrylate	Acrylonitrile	Na naphthalene	THF	BAB	105, 106
		BuLi/sulfur-containing modifier	–	AB	107
		KNH_2/2-aminoethanol	–	AB	108, 109
	Methacrylonitrile	BuLi/sulfur-containing modifier	–	BAB, AB	107, 110
			–	ABA	111
Methyl acrylate	Acetaldehyde	BuLi	Toluene	AB	23
		Alkali metal–aromatic complex	Toluene/THF	BAB	23
Methyl methacrylate	Formaldehyde	9-Fluorenyllithium	Toluene/ethyl ether	AB	5, 7
		BuLi	Toluene	AB	23
		Alkali metal–aromatic complex	THF/toluene	BAB	23
		BuLi	Benzene	AB	6, 8
	Caprolactone	Na α-methylstyrene	–	BAB	93
	n-Butyl isocyanate	Na biphenyl	Toluene (10% THF)	BAB	3
	Toluene diisocyanate	*n*-BuLi	THF/toluene	AB	100
	Butyl methacrylate	Na naphthalene	THF	BAB	105
	Hexyl methacrylate	Na naphthalene	THF	BAB	105
		Φ_2CHNa or Φ_2CHK	THF	AB	112
	Isopropyl acrylate	Na naphthalene	THF	BAB	105
	Octadecyl methacrylate	Φ_2CHNa or Φ_2CHK	THF	AB	112
	1-Methacryloxy-2-butyne	Na naphthalene	THF	BAB	113
	Ethylene sulfide	BuLi	THF	AB	114
		Na naphthalene	THF	ABA	114

	Propylene sulfide	BuLi	THF	AB	114
		Na naphthalene	THF	BAB	114
	Lauryl methacrylate	Φ_2CHNa or Φ_2CHK	THF	AB	112
		BuLi	Toluene/THF	ABA	115
	Isoprene	*s*-BuLi/DVB	Toluene/THF	ABA	116
		Li naphthalene	THF	ABA	97
		K naphthalene	THF	ABA	117
	Hexamethylcyclotri-siloxane	BuLi	THF/$(Me_2N)_3PO$	AB	118
		Li benzophenone	THF/$(Me_2N)_3PO$	AB	118
		Dilithiobenzophenone	THF/$(Me_2N)_3PO$	BAB	119
		Dilithiothiobenzophenone	THF/$(Me_2N)_3PO$	BAB	119
		Dilithio-*N*-(diphenylmethylene) aniline	THF/$(Me_2N)_3PO$	BAB	119
		Dilithiofluorenone	THF/$(Me_2N)_3PO$	BAB	119
	Octamethylcyclotetra-siloxane	BuLi	THF	AB	18
		Li (or Na or K) naphthalene	THF	BAB	18
	Thiocarbonyl fluoride	Na biphenyl	THF	BAB	19, 20
	Hexafluoroacetone	Na biphenyl	THF	BAB	19
Butyl methacrylate	Formaldehyde	9-Fluorenyllithium	Toluene–diethyl ether	AB	5–7
		BuLi	Benzene	AB	8
Ethyl methacrylate	Ethylene oxide	Diphenyl methylpotassium	THF	AB	120
Hexyl methacrylate	Lauryl methacrylate	Φ_2CHNa or Φ_2CHK	THF	AB	112
Isopropyl acrylate	Acrylonitrile	Na naphthalene	THF	BAB	105
Styrene	Acetaldehyde	BuLi	Toluene	AB	23
		Na naphthalene	Toluene/THF	BAB	23
	Butyraldehyde	BuLi	Toluene	AB	23
		Na naphthalene	Toluene/THF	BAB	23

(*table continues*)

TABLE 9.1 Block Copolymers – Two Monomers (*continued*)

Monomer A	Monomer B	Initiator	Solvent	Block sequence	References
Styrene (*continued*)	Formaldehyde	Na naphthalene	THF	BAB	4, 5, 7, 9
		BuLi	Benzene	AB	5–9
		BuLi	Toluene	AB	23
		Na biphenyl	THF	BAB	24
		Alkali metal–aromatic complex	Toluene/THF	BAB	23
		BuNa or ΦLi	Benzene	AB	7, 9
		Li naphthalene	THF	BAB	121
	Hexafluoroacetone	Na biphenyl	THF	BAB	19
	Thiocarbonyl fluoride	Na biphenyl	THF	BAB	19, 20
	Acrylonitrile	Na naphthalene	THF	BAB	105, 106, 122
	Methacrylonitrile	Phenylisopropylpotassium	THF	AB, BAB	96
		Na α-methylstyrene			
		Na naphthalene	THF	BAB	105, 123
	Butadiene	Na α-methylstyrene	THF	ABA, BAB	124
		n-BuLi	Benzene/cyclohexane	ABA	3
		s-BuLi	Cyclohexane	ABA, ABABA	125–127
		Aromatic ketone dilithium complex	Toluene	ABA	128
		Li methylnaphthalene	Ether–cyclohexane	ABA	129, 130
		1,2-Dilithio-1,2-diphenylethane	Ether, cyclohexane	ABA	131, 132
		Dilithiobutane	Ether/toluene	ABA	133
		Dilithioisoprene	Benzene (triether)	ABA	134, 135
		s-BuLi	Benzene	ABA	136–142
		n-BuLi	Benzene	AB, ABA, BAB, ABAB	143
		EtLi	Benzene/anisole or diphenyl ether	ABA	144–146
		BuLi	Toluene	ABA	147

	s-BuLi	Cyclohexane	AB, ABA, AB > BA AB AB BA × AB BA	148
	Propyllithium	Toluene	ABA	149
	BuLi	Cyclohexane	ABA	150
	BuLi	Toluene	AB	151, 152
	BuLi	Benzene	ABA	138, 153–156
	s-BuLi	Benzene	AB	135
	BuLi	–	AB BA × AB BA	157
	s-BuLi	Cyclohexane	AB	25, 158, 159
	BuLi	n-Hexane	AB	160–162
	BuLi	Benzene	AB	153, 163, 164
	BuLi	No solvent	AB	165, 166
	BuLi	Heptane	AB	104
	BuLi	Toluene/THF	AB	167
	1-Hydroxy-2-aminoethane/KNH_2 complex	Liq NH_3	AB	162
	s-BuLi	Isopentane/cyclohexane	ABA	168
	BuLi/*trans*-stilbene	Benzene	ABA	169
	K α-methylstyrene	THF	BAB	170
Isoprene	Li metal	Benzene	ABA	171
	Li metal	Dioxane	ABA	172
	Dilithioisoprene	Benzene (triether)	ABA	134
	n-BuLi	Benzene/cyclohexane	ABA	3, 173
	s-BuLi	Isopentane/cyclohexane	ABA	140–142 169, 174

(*table continues*)

TABLE 9.1 Block Copolymers – Two Monomers (*continued*)

Monomer A	Monomer B	Initiator	Solvent	Block sequence	References
Styrene (*continued*)	Isoprene (*continued*)	*s*-BuLi	Benzene	ABA	124, 136, 138, 175–178
		Na naphthalene	THF	ABA, BAB	179–181
		Na naphthalene	THF	$A(BAB)_2$	179–181
		Na naphthalene	THF	$B(ABAB)_2$	179–181
		n-BuLi	*n*-Hexane/benzene	AB	96, 182
		Phenylisopropylpotassium	THF	AB	96
		Na α-methylstyrene	THF	BAB	96
		Cs naphthalene	THF	ABA	183
		Na α-methylstyrene	THF	ABA, BAB	184
		n-BuLi or *s*-BuLi	Cyclohexane	ABA	185
		1,4-Dilithio-1,1,4,4-tetraphenylbutane	Cyclohexane/THF	ABA	186
		EtLi	Benzene/anisole	ABA	144–146
		n-BuLi	Cyclohexane	AB	25, 187
		s-BuLi	Cyclohexane	ABA	188
		n-BuLi	Benzene	AB	189, 190
		s-BuLi	Benzene	$(AB)_3$	191
		s-BuLi	Benzene	AB	192
		n-BuLi	THF	AB	193, 194
		n-BuLi	Hexane or benzene	ABA	138, 195
		n-BuLi	Benzene	ABA, AB∨BA (AB over BA), AB×BA (BA, AB over AB, BA)	196
		1,4-Dilithiobutane	THF	ABA	194
		s-BuLi	–	AB	165
		Li naphthalene	THF	ABA	97
		s-BuLi	Cyclohexane	AB	158
		n-BuLi	Toluene	AB, $(AB)_2$, ABA	189, 197

ε-Caprolactam	BuLi	Hexane	AB	28
	Polystyrene bischloroformate	THF	BAB	198
α-Pyrrolidone	*N*-acylpyrrolidone-ended polystyrene (difunctional)/NaH or Na naphthalene	Bulk	BAB	199
	Polystyrene bischloroformates	THF	BAB	198
Carbobenzoxy-L-lysine NCA	Cumylpotassium	THF	AB	34
L-Lysine NCA	Cumylpotassium	THF	AB	32
n-Butylisocyanate	Na biphenyl	Toluene (2% THF)	BAB	200
	n-BuLi	Toluene/THF	AB	100
	Na naphthalene	THF	BAB	201
Toluene diisocyanate	*n*-BuLi	Toluene/THF	AB	100
Caprolactone	BuLi	–	AB	29, 201
	Na	THF/benzene	BAB	201
Pivalolactone	*n*-BuLi	Cyclohexane	AB	202
			AB	203
	Na α-methylstyrene	THF/Me_2SO	BAB	17, 203, 204
Allyl methacrylate	Na naphthalene	THF	BAB	205
Hexyl methacrylate	Na naphthalene	THF	BAB	105
Methyl methacrylate	Na biphenyl	THF	BAB	206–208
	BuLi	Toluene	AB	209
	Li metal, Li phenoxide	THF	AB	210
	Li β-naphthoxide	THF	AB	210
	Na naphthalene	THF	BAB	106, 211, 212
	Na α-methylstyrene	THF	BAB	96, 213
	Na naphthalene	DME	BAB	214
	Phenylisopropylpotassium	THF	AB	96, 215–217
	n-BuLi	*n*-Hexane	AB	208
1-Methacryloxy-2-butyne	Na naphthalene	THF	BAB	113

(*table continues*)

TABLE 9.1 Block Copolymers – Two Monomers (*continued*)

Monomer A	Monomer B	Initiator	Solvent	Block sequence	References
Styrene (*continued*)	Ethylene oxide	Na naphthalene	DME	BAB	214
		Na α-methylstyrene	THF	BAB	96, 218–220
		Phenylisopropylpotassium	THF	BAB	96
		K biphenyl	THF, dioxane, and MeOH	BAB	221
		Cumylpotassium	THF	AB	219, 222–224
		K α-methylstyrene	THF	BAB	223
		Na naphthalene	THF	BAB	225
	Tetrafluoroethylene oxide	Na biphenyl	THF	BAB	19, 20
	Propylene oxide	Na α-methylstyrene	THF	BAB	220
	Butylene oxide	Na α-methylstyrene	THF	BAB	220
	Ethylene sulfide	BuLi	THF	AB	114, 226
		Na naphthalene	THF	BAB	114, 227, 228
		K naphthalene	THF	BAB	229
		Na biphenyl	THF	BAB	230
	Tetrafluorcethylene sulfide	Na biphenyl	THF	BAB	19, 20
	Propylene sulfide	BuLi	THF	AB	114
		Na naphthalene	THF	BAB	114, 231, 232
		Na biphenyl	THF	BAB	86
	Hexamethylcyclo-trisiloxane	K naphthalene	THF	BAB	233
		Li naphthalene	THF/toluene	BAB, $(BAB)_{2-10}$	234
		BuLi	Hexane/THF or glyme	AB	235, 236
		s-BuLi	Cyclohexane/diglyme	AB	101
		Dilithium stilbene	Benzene/THF	BAB	237
		Li naphthalene	THF/toluene	BAB	238, 239

	Li ΦCHCHΦLi	THF or diglyme	$(BAB)_n$	236, 240
	s-BuLi	Benzene/THF	AB	241
	BuLi	?	ABA	242
	BuLi	Cyclohexane/THF	AB	243, 244
	BuLi	THF	ABA	245, 246
	BuLi	THF/benzene	AB, ABA	247
	s-BuLi	THF/benzene	ABA	248
Octamethylcyclotetra-siloxane	K metal, K naphthalene	THF	BAB	102
	BuLi	THF	AB	18
	K (or Na, Li) naphthalene	THF	BAB	18
2,4,6-tris(3,3,3-trifluoropropyl)-2,4,6-trimethylcyclotrisiloxane	BuLi	THF	AB	249
$ClSi(Me)_2[OSi(Me)_2]_nCl$	BuLi	Benzene–anisole	ABA	250
	Na naphthalene	THF	$(AB)_n$	251
Piperylene	BuLi	Heptane	BA	104
α-Methylstyrene	Na naphthalene	THF	BAB	214
	s-BuLi	THF	AB	252, 253
	Dilithiopolybutadiene oligomer	Benzene/$(Me_2N)_3PO$	ABA	254
	Cumylpotassium	THF	AB	255
	Na naphthalene	THF	ABA, BAB	256
α-Vinylnaphthalene	Phenylisopropylpotassium	THF	AB, BAB	96
	Na α-methylstyrene	THF	AB, BAB	96
2-Vinylpyridine	Na naphthalene	THF	BAB	257
	Phenylisopropylpotassium	THF	AB	96
	Na α-methylstyrene	THF	BAB	96
	Cumylpotassium	THF	AB	258–260
	Na biphenyl	THF	BAB	260

(table continues)

TABLE 9.1 Block Copolymers – Two Monomers (*continued*)

Monomer A	Monomer B	Initiator	Solvent	Block sequence	References
Styrene (*continued*)	2-Vinylpyridine (*continued*)	BuLi	THF	AB, ABA, BAB	261
		Na α-methylstyrene	THF	$(AB)_n$	261
	4-Vinylpyridine	Phenylisopropylpotassium	THF	AB	
		Na α-methylstyrene	THF	BAB	
		Cumylpotassium	THF	AB	259
	1-Vinylpyrene	BuLi	Benzene	AB	261
		Cumylpotassium	THF	AB	261
		Alkali metals–α-methylstyrenes	THF	ABA	261
	3-Vinylpyrene	BuLi	Toluene/decalin	AB	261
	p-Bromostyrene	Phenylisopropylpotassium	THF	AB, BAB	96
		Na α-methylstyrene	THF	BAB	96
	t-Butylstyrene	BuLi	Benzene	ABA	262
	p-Chlorostyrene	Na naphthalene	THF	ABA, BAB	263
	Ring-substituted styrenes 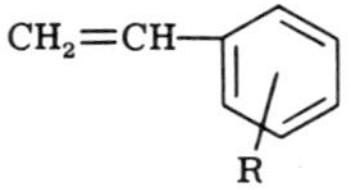	Na naphthalene	THF	BAB	264
	1,2-Dihydro-naphthalene	Na naphthalene	THF	BAB	257
p-Bromostyrene	Ethylene oxide	Na α-methylstyrene	THF	BAB	96
		Phenylisopropylpotassium	THF	AB	96
	Methyl methacrylate	Na α-methylstyrene	THF	BAB	96
		Phenylisopropylpotassium	THF	AB	96
α-Methylstyrene	Acetaldehyde	BuLi	Toluene	AB	23
		Alkali metal–aromatic complex	Toluene/THF	BAB	23
	Formaldehyde	Na naphthalene	THF	BAB	4, 5, 7
		BuLi	Toluene	AB	23

	Alkali metal–aromatic complex	Toluene/THF	BAB	23
	BuLi	Benzene	AB	8
Ethylene oxide	Na α-methylstyrene	THF	BAB	265
Ethylene sulfide	BuLi	THF	AB	265
	Na naphthalene	THF	BAB	265
Propylene sulfide	BuLi	THF	AB	265
	Na naphthalene	THF	BAB	265
	EtLi	THF	ABA	266
n-Butyl isocyanate	Na biphenyl	Toluene (10% THF)	BAB	3
	n-BuLi	Toluene	AB	3
Butadiene	Na dispersion	THF	BAB	267, 268
	s-BuLi	Nonpolar solvents	ABA	269
	BuLi	Cyclohexane/THF	ABA	270
	Dilithium diisoprene	Benzene/hexane	ABA	139, 269
	1,2-Dilithio-1,2-diphenylethane	Toluene/$(Me_2N)_3PO$	ABA	271
(Diene)	Organodilithium catalyst	Hydrocarbon solvents	ABA	272
Butadiene	BuLi	Benzene/hexane/dimethoxyethane	ABA	139
	s-BuLi	THF/toluene	ABA	273
Butadiene or isoprene	1,4-Dilithio-1,1,4,4-tetraphenylbutane	*n*-Heptane/TMEDA	ABA	274
Butadiene	BuLi	THF	AB	275
Isoprene	1,4-Dilithio-1,1,4,4-tetraphenylbutane	Cyclohexane/THF	ABA	140, 142 145, 186
	Dilithio-1,4-dimethyl-1,4-diphenylbutane	Toluene/trace THF	ABA	276
	Dilithium polyisoprene	Nonpolar solvents	ABA	269
	Dilithiobutane	Ether	ABA	136
	Li naphthalene	THF	ABA	97
	s-BuLi	Isopentane/cyclohexane	ABA	169

(*table continues*)

TABLE 9.1 Block Copolymers – Two Monomers (*continued*)

Monomer A	Monomer B	Initiator	Solvent	Block sequence	References
α-Methylstyrene (*continued*)	Isoprene (*continued*)	1,2-Dilithio-1,2-diphenylethane/ $(Me_2N)_3PO$	Toluene	ABA	271
		s-BuLi	Pentamethyl- heptane/THF	AB	277
α-Methylstyrene	Me methacrylate	Na biphenyl	THF	BAB	278, 279
α-Methylstyrene	α,ω-Dihalopoly- methylene	Na naphthalene	THF	$(AB)_x$	280
	Hexamethylcyclo- triloxane	*s*-BuLi	THF/benzene	ABA	248
	Octamethylcyclotetra- siloxane	Na–K alloy	THF	BAB	281
α-Methylstyrene/ styrene	Isoprene	*s*-BuLi	Pentamethyl- heptane/THF	AB	277
	Butadiene	*s*-BuLi	THF	$AB(B\text{–}A)_{1-10}$	282
t-Butylstyrene	Butadiene	*s*-BuLi	Cyclohexane	AB, ABA	283
2,4-Dimethyl- styrene	Isoprene	BuLi	Hexane	AB	284
	Butadiene	BuLi	Hexane	AB	284
1,1-Diphenyl- ethylene/styrene	Butadiene	*s*-BuLi	Cyclohexane	AB, ABA	285
2-Vinylpyridine	Propylene sulfide	Na naphthalene	THF	BAB	40
		Na biphenyl	THF	BAB	40, 286
		Φ_2CHNa	THF	AB	287
		Φ_2CHNa	THF	BAB	286
	t-Butyl acrylate	BuLi	THF	AB	288
	Isopropyl acrylate	BuLi	THF	AB	288
	Methyl methacrylate	BuLi	THF	AB	288
	Trimethylsilyl methacrylate	BuLi	THF	AB	288

	Hexamethylcyclotri-siloxane	Li naphthalene	Toluene/THF	BAB	289, 290
	4-Vinylpyridine	Diphenylmethylsodium	THF	AB	291
4-Vinylpyridine	Methyl methacrylate	Phenylisopropylpotassium	THF	AB	96
		Na α-methylstyrene	THF	BAB	96
2-Vinylpyridine	3-Vinylpyrene	BuLi	Toluene	AB	21
4-Vinylpyridine	Ethylene sulfide	Na naphthalene	THF	BAB	228
	Hexaphenylcyclotri-siloxane	Li naphthalene	Toluene/THF	BAB	290
	p-Bromostyrene	Phenylisopropylpotassium	THF	AB	96
		Na α-methylstyrene	THF	BAB	96
2-Methyl-5-Vinyl-pyridine	*n*-Butyraldehyde	BuLi		AB	23
	Isobutyraldehyde	BuLi		AB	23
	Formaldehyde	Na naphthalene	THF	BAB	4
		BuLi	Toluene	AB	23
		Alkali metal–aromatic complex	Toluene/THF	BAB	23
4-Vinylbiphenyl	Isoprene	BuLi	Benzene	ABA	292
Vinyltriorgano-silane	Isoprene (or butadiene)	*s*-BuLi	Cyclohexane	ABA	293
	Isoprene	*s*-BuLi	Cyclohexane	ABA	294

[a] Does not include star-branched block copolymers, which can be found in Table 10.2.

[b] (A-co-B) represents a copolymer of butadiene and diphenylethylene.

[c] Poly(ethylene oxide)glycol.

TABLE 9.2 Block Copolymers by Anionic Polymerization of Three Monomers

Monomer A	Monomer B	Monomer C	Initiator	Solvent	Block sequence	References
Butadiene	Isoprene	Styrene	s-BuLi	Isopentane	ABC	295
			Na α-methylstyrene	THF	BCACB	296
	Styrene	2-Vinylpyridine	1,4-Dilithio-1,1,4,4-tetraphenylbutane	Benzene/Et_2O/toluene	CBABC	297
			s-BuLi	Cyclohexane	BABC	298
Ethylene oxide	1,1-Dimethyl-ethylene oxide	Propylene sulfide	9-Carbazylpotassium	THF	ABC	73
Hexamethylcyclo-trisiloxane	$[Si(CH_3)\Phi O]_3$	$(Si\Phi_2O)_3$	$(LiOSi\Phi_2)_2O$ (Φ above and below Si)	THF/benzene	CBABC	299
			$LiOSi\Phi_2OLi$ (Φ above and below Si)	THF/anisole	ACBCA BCACB	300
Styrene	Butadiene	Acrylonitrile	Organoalkali	–	ABC	301
		ε-Caprolactam	s-BuLi	Cyclohexane/decalin	ABC	302
	Isoprene	Acrylonitrile	s-BuLi	Cyclohexane	ABC	303
		Ethylene sulfide	n-BuLi	Benzene/anisole	ABC	304
			n-BuLi/anisole	Cyclohexane/THF	ABC	305
	Hexamethyl-cyclotrisiloxane	Hexaphenylcyclo-trisiloxane	BuLi	Cyclohexane/THF	ABC	306
	2-Vinylpyridine	Methyl methacrylate	Na	THF	CBABC	288
α-Methylstyrene	α-Vinylpyridine	Methyl methacrylate	Na	THF	CBABC	288

monomers which are amenable to this mechanism. Table 9.1 lists those block copolymers which contain two or more blocks obtained from two different monomers, while Table 9.2 comprises a list of those block copolymers prepared from three different monomers. As indicated in the prior paragraphs, any impurities in the system could act to terminate one or more of the blocks, leading to inclusion of "free" blocks in the final material, and this could affect the physical behavior of the latter. In most cases listed in the above tables, there were no available data on the purity of the block copolymers. Hence, as in the case of previous data collated from the open literature, the reader would have to use considerable judgment in accepting any correlation of the mechanical properties of the materials to their alleged structure.

3. Morphology and Structure–Property Relations of ABA Triblock Copolymers*

As indicated earlier, the triblock copolymers of the ABA type have proven to be the most interesting scientifically and the most important technologically. It should be noted at this point that, on the basis of their morphology and mechanical behavior, this class of polymers also includes the "polyblock" copolymers of the $A(BA)_x$ or $(AB)_x$ types as well. In other words, the *minimum* sequence required is ABA, so that the A blocks occur at *each* end of the B blocks.

a. Morphology of Styrene–Diene–Styrene Triblock Copolymers

When this type of "molecular architecture" includes A blocks of a glassy (or crystalline) polymer (e.g., polystyrene) and B blocks of a rubbery polymer (e.g., polybutadiene or polyisoprene), an interesting product is obtained, i.e., a "thermoplastic elastomer." The most popular variety has been the polystyrene–polybutadiene–polystyrene triblock copolymer, and a schematic of its structure and morphology is shown in Fig. 9.1. It can be seen that this type of polymer leads to a heterogeneous material consisting of polystyrene "domains" dispersed within a matrix of polybutadiene. This is due, of course, to the thermodynamic incompatibility of the two types of blocks at the indicated molecular weights, since even a small positive heat of mixing is sufficient to exceed the very small entropy of mixing of polymers.

* See Morton.[140,142]

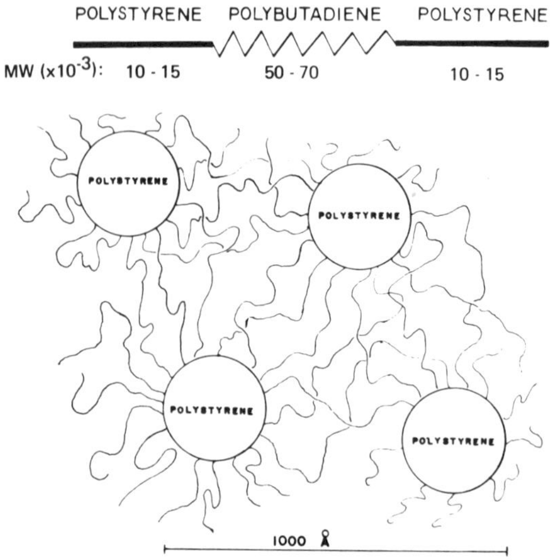

Fig. 9.1 Schematic of a styrene–butadiene–styrene triblock copolymer.[140]

The morphology depicted in Fig. 9.1 therefore represents a network structure, where the small polystyrene domains act as network junctions for the elastic polybutadiene chains. Thus this material behaves in an analogous fashion to a crosslinked (vulcanized) rubber at room temperature, but is "thermoplastic," i.e., can be easily made to flow, at elevated temperatures (100–140°C) where the polystyrene domains soften and liquefy. Hence the name "thermoplastic elastomers."

However, it should be pointed out that the molecular architecture of these materials must be within the relatively narrow limits shown in Fig. 9.1 in order to attain the optimum mechanical properties. This will become clear in the subsequent discussion, but suffice it to say that the lower limit in block molecular weight is set by the minimum chain length of polystyrene required for phase separation, while the upper limit is governed by viscosity considerations which can both interfere with good phase separation and restrict the flow during molding. It should be noted, too, that optimum properties also require a polystyrene content of about 30%, since that is close to the maximum volume fraction consistent with having the polybutadiene as the matrix and thus retaining the rubbery properties of the composite.

The morphology proposed in Fig. 9.1 has been verified by direct observation with electron microscopy,[143,307–310] using mainly stained thin films prepared from molded or solvent-cast films, as shown in a typical case in Fig. 9.2. Small-angle X-ray scattering[311] has also indicated the presence of spherical domains of polystyrene in an orderly array (a "macrolattice"), as shown in Fig. 9.3, where the polystyrene spheres have a diameter of 35.6 nm and the orthorhombic unit cell has the dimensions of 67.6, 67.6, and 56.6 nm. The latter dimensions were obtained for a polystyrene–polybutadiene–polystyrene (SBS) triblock of molecular weight 21,100–63,400–21,100.

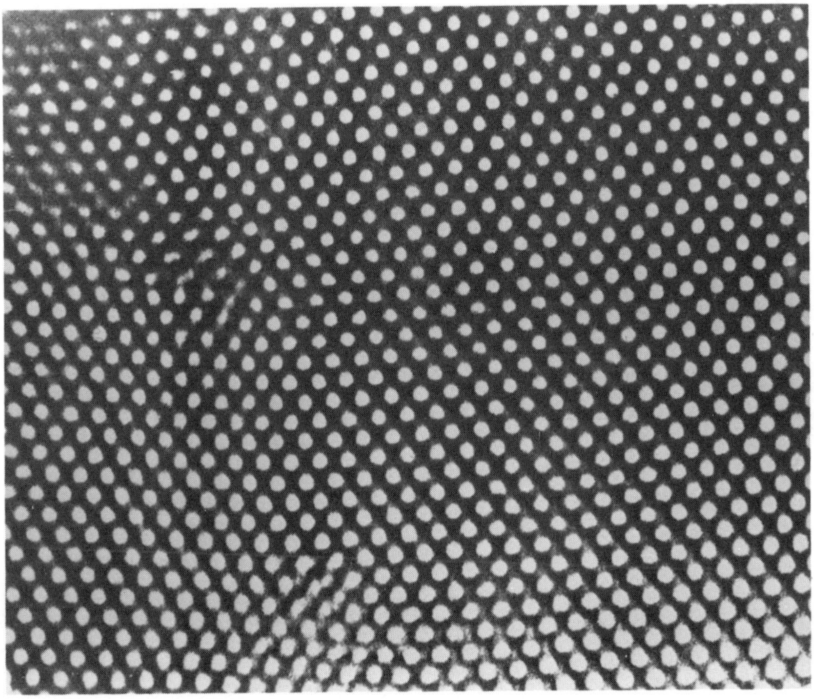

Fig. 9.2 Electron microphotograph of a styrene–isoprene–styrene triblock copolymer (MW 16,200–75,600–16,200). Ultrathin (~ 50 nm) section of film cast from 90/10 H_4-furan/butanone and stained with OsO_4 (X 85,000).

The small size of the polystyrene domains, shown as white circles in the OsO_4-stained film of Fig. 9.2, accounts for the fact that these elastomers are virtually transparent to visible light. Their size and shape are, of course, dependent on the molecular architecture, the aggregation of the polystyrene blocks into separate domains being restricted by the limitation imposed by the end-to-end distance of the central polybutadiene block. Since the latter also have the characteristic of remarkable uniformity in size, it is not surprising that the interdomain distances are also remarkably uniform. The formation of the polystyrene domains may, of course, also be governed by the interfacial energy[311] and by the kinetics of the phase separation process, which could be markedly affected by the viscosity.

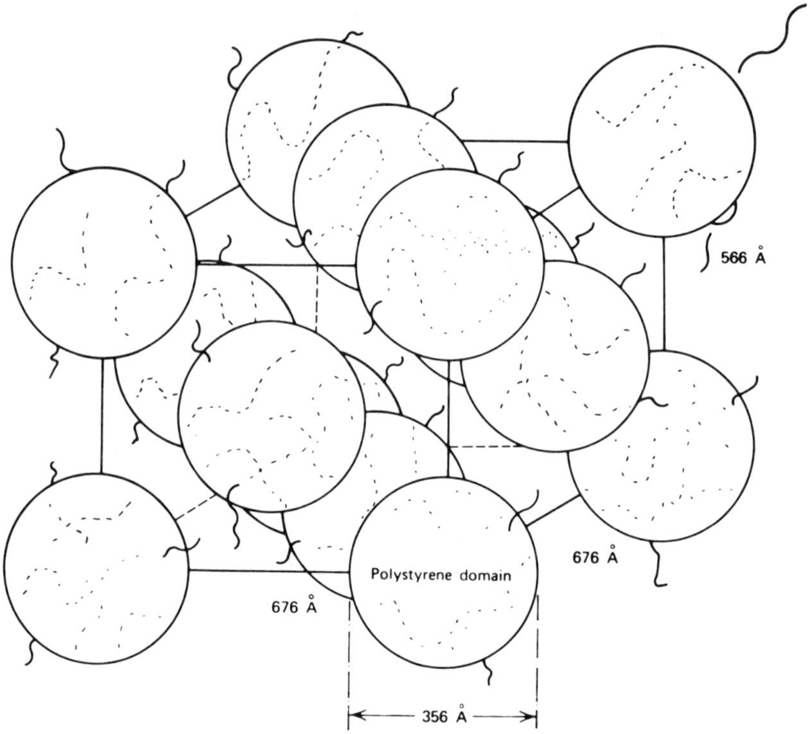

Fig. 9.3 Scale model of macrolattice of a styrene–butadiene–styrene triblock copolymer (21,100–63,400–21,100) calculated from low angle X-ray scattering.[140,311] (Reprinted with permission from Ref. 311. Copyright 1970 American Chemical Society.)

The clear-cut and uniform spherical domains shown in Figs. 9.2 and 9.3 are a direct result of both the synthesis of the polymer and the method of preparation of the sample. Thus the domain uniformity is undoubtedly due to the virtual monodispersity of molecular weights of the blocks and of the overall polymer. Typical[140,144] gel permeation chromatograms of a styrene–butadiene–styrene (SBS) and a styrene–isoprene–styrene (SIS) block copolymer prepared under rigorous high-vacuum conditions are shown in Fig. 9.4, and demonstrate the uniformity of the molecular weights and the absence of monoblock or diblock "impurities." The specimens used for Figs. 9.2 and 9.3 were cast from solution and annealed in order to come as close as possible to an equilibrium in the phase separation. Under less ideal conditions, e.g., for molded or extruded specimens, or where the polystyrene blocks exceed about 30% of the total volume, the domains may assume forms other than spherical, e.g., cylindrical or lamellar.[312] Such morphologies may introduce some anisotropy in the mechanical behavior.

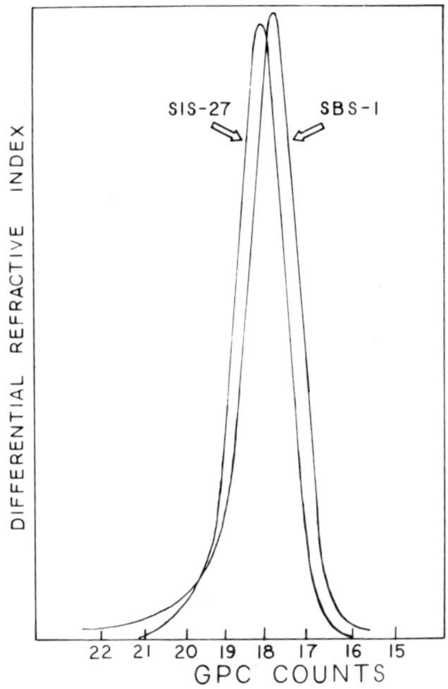

Fig. 9.4 Gel permeation chromatograms of styrene–butadiene–styrene (SBS) and styrene–isoprene–styrene (SIS) triblock copolymers.[140] (Courtesy of John Wiley and Sons.)

b. Structure–Property Relations of Elastomeric Triblock Copolymers

Because of their technological importance, the styrene–diene triblock copolymers which result in thermoplastic elastomers have undoubtedly received the greatest attention. Thus there is now a substantial body of knowledge[140,142] about the factors which affect their mechanical properties, with special reference to their stress–strain behavior and tensile strength. It should be remembered that such information is based on studies carried out on well-characterized, very "pure" polymers,[140–146] and that these structure–property relations may be obscured in the case of commercial materials.

This type of thermoplastic elastomer can be considered as consisting of an elastic network of polydiene chains held together by the polystyrene domains, which act both as network junctions and, because of their substantial proportion, as a finely divided filler. (The particle size of this filler is even smaller than that of most carbon blacks, and hence one can expect a substantial reinforcement of mechanical properties.) Hence both the block size and proportion of polystyrene might be expected to have an influence on such properties.

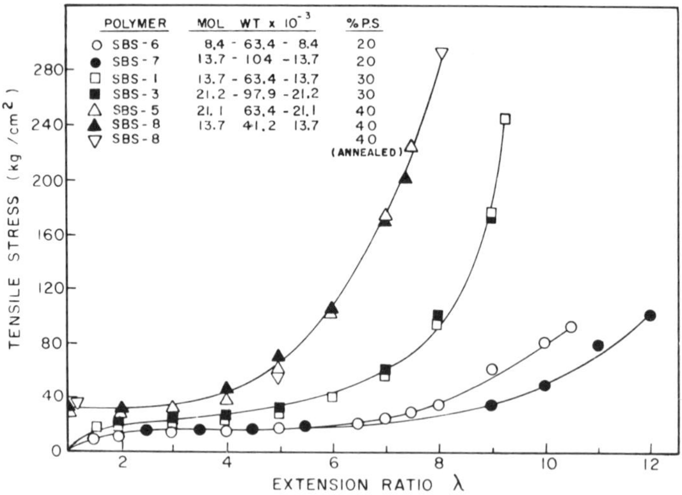

Fig. 9.5 Effect of composition and block size on tensile properties of styrene–butadiene–styrene triblock copolymers (molded sheets).[140] (Courtesy of John Wiley and Sons.)

A series of styrene–butadiene–styrene (SBS) polymers which illustrate these effects is shown in Fig. 9.5. It can be seen that *both* the stress levels and the ultimate stress (tensile strength) are a monotonic function of the polystyrene block content, but that the molecular weight of the blocks has no influence, within the limits shown. This vitiates any idea that the polybutadiene chains represent the "molecular weight between crosslinks" (M_c), and this is not surprising since these chains are of a size to include many *entanglements* which could act as virtual network junctions.[313] On the other hand, the polystyrene, acting as a rigid filler, would be expected to influence both the stress levels and the strength in a monotonic fashion.

It should be noted that the specimens used in this work had been molded (10 min at 140°C) and that such a process may not permit an approach to equilibrium in phase separation. This is especially noticeable in the case of the 40% polystyrene content polymer in Fig. 9.5, where annealing (i.e., slow cooling of the molded sample at 1°C/min compared to rapid cooling at 20°C/min) resulted in an increase in tensile strength (but not in modulus level). This annealing effect was not observed at lower polystyrene content, and illustrates the importance of considering the processing history of these materials in evaluating their properties. It was actually found in this work that films of these materials, cast from suitable solvents, gave the best and most reproducible results.[140]

Another interesting feature of Fig. 9.5 is the marked difference in the curves at *low strains*, where the 40% polystyrene curves are the only ones to show a

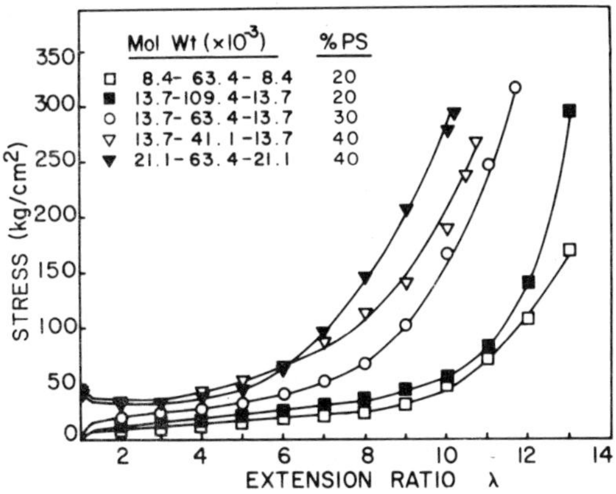

Fig. 9.6 Effect of composition and block size on tensile properties of styrene–isoprene–styrene block copolymers (cast films).[140] (Courtesy of John Wiley and Sons.)

distinct yield point. This behavior was not reversible, since it occurred only with the *first draw* and not on subsequent extensions. However, it could be reproduced if the samples were remolded or reheated. This behavior is ascribed to the fact that, at this high level of styrene content, the polystyrene domains may reach or exceed the critical volume fraction for packing, so that they would form a continuous interpenetrating phase. Such "connections" between the polystyrene domains would be expected to be broken upon extension, and should not reform unless the material is reheated.

Stress–strain curves for a similar series of styrene–isoprene–styrene (SIS) triblock copolymers are shown in Fig. 9.6, and show characteristics generally similar to the SBS curves in Fig. 9.5, except for one difference: there does not appear to be any dependence of tensile strength on the polymer architecture. The only SIS polymer which shows a substantially lower strength is the one containing 20% polystyrene with the low block molecular weight of 8400. Since these polymers do not show a dependence of strength on polystyrene *content*, the theory proposed above that the polystyrene domains act as a strength-reinforcing filler does not seem valid. Instead, it appears that one must examine the factors which contribute to the *integrity* of the polystyrene domains, since it is they which hold the network together. This, then, indicates that the *degree of phase separation* might provide the key to the behavior of these elastomers.

The solubility parameters[314] for polystyrene ($\delta = 8.83$), polybutadiene ($\delta = 8.44$), and polyisoprene ($\delta = 8.10$) indicate that phase separation should be better in the SIS than in the SBS polymers. Since such phase separation is also a function of molecular weight and proportion of the two components,[315] this would account for the increase in tensile strength of the SBS polymers

TABLE 9.3 Stress–Strain Properties of SIS Triblock Copolymers[a]

Wt % styrene	Molecular weight ($\times 10^{-3}$)			Tensile strength (MPa)	Stress at $\lambda = 4$ (MPa)
	S	I	S		
40	21.1	63.4	21.1	31.0	3.7
40	13.7	41.1	13.7	30.6	4.3
30	13.7	63.4	13.7	32.1	2.4
20	13.7	109.4	13.7	27.0	1.8
20	8.4	63.4	8.4	16.0	1.1
19	7.0	60.0	7.0	2.2	1.3
11	5.0	80.0	5.0	~0	~0

[a] From Morton[140,142] and Morton and Goh.[141]

(Fig. 9.5) with increasing styrene content. In the case of the SIS polymers on the other hand, the degree of phase separation has apparently already reached its maximum except in the case of the polymer having polystyrene end blocks of 8400 molecular weight, where phase mixing is apparently still serious. The effect of polystyrene content and molecular weight on the stress–strain properties of these SIS polymers is summarized in Table 9.3, which contains some of the data from Fig. 9.6. It is clear from these data that, while the modulus of these polymers (stress values) depends primarily on the polystyrene content (filler effect), the tensile strength becomes very sensitive to the molecular weight of the polystyrene at the *lower* values. It also appears that a polystyrene molecular weight of about 6000 represents a critical value for any phase separation, since lower values lead to material of no measurable integrity. It is interesting that this corroborates theoretical predictions[315,316] based on thermodynamic calculations.

Further confirmation of the critical role of the polystyrene domains in controlling the elastic and strength properties is provided by studies on an analogous triblock copolymer also having a polyisoprene center block but end blocks of poly-α-methylstyrene.[140,142,186] The stress–strain properties of this polymer are shown in Fig. 9.7, while the effect of temperature on the tensile strength of this polymer is compared with that of the SIS polymer in Fig. 9.8. It is obvious at once that the higher T_g of the poly-α-methylstyrene blocks increases the resistance of this polymer to the effect of higher temperatures (Fig. 9.8). However, it is also interesting to note that the stress at any given strain is also higher for the α-methylstyrene (Fig. 9.7), presumably due to the higher modulus of the poly-α-methylstyrene domains. This result confirms the fact that the glassy domains in these types of materials are not entirely rigid, but are considerably distorted under stress, especially at high strains, as actually observed by electron microscopy of highly strained specimens.[307,308] They thus act as "energy sinks," absorbing the elastic energy which would

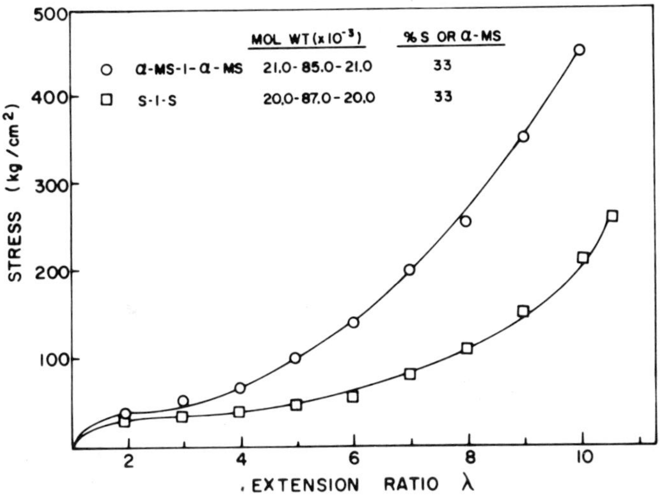

Fig. 9.7 Tensile properties of an α-methylstyrene–isoprene–α-methylstyrene triblock copolymer compared to a similar styrene–isoprene–styrene polymer.[140] (Courtesy, John Wiley and Sons.)

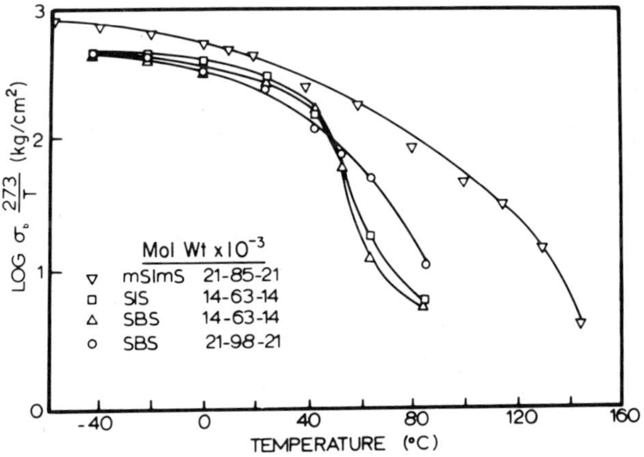

Fig. 9.8 Effect of temperature on tensile strength of α-methylstyrene and styrene triblock copolymers (S, styrene; mS, α-methylstyrene; B, butadiene; I, isoprene).[140] (Courtesy, John Wiley and Sons.)

otherwise lead to rupture.[317] This accounts for the very high strengths exhibited by these materials, but also unfortunately leads to high degrees of unrecoverable deformation ("set").[145]

c. "STAR"-BRANCHED TRIBLOCK COPOLYMERS

A special variation of the styrene–diene ABA triblock copolymers that deserves mention is the class known as "star"-branched or "radial" block copolymers.[318–320] These are prepared by first synthesizing an AB diblock by the usual organolithium polymerization, followed by a "linking" reaction by means of a multifunctional agent such as the chlorosilanes or divinylbenzene. The reaction may be shown schematically as

$$RLi + xS \rightarrow RS_xLi \tag{9.1}$$

$$RS_xLi + yI \rightarrow RS_xI_yLi \tag{9.2}$$

$$zRS_xI_yLi + R'X_z \rightarrow R'(RS_xI_y)_z + zLiX \tag{9.3}$$

where S = styrene, I = isoprene, and $R'X_z$ = a z-functional linking agent. Hence, although the chain structure of this type of polymer is branched, with all the branches emanating from a central point, the *sequence* of blocks still follows the ABA pattern, with the polystyrene still comprising the A end blocks.

Block copolymers of styrene and isoprene with up to 30 branches have been prepared in this way, using divinylbenzene as linking agent.[318,320] Because of their branched structure, these polymers exhibit both different morphologies and different mechanical properties than their linear counterparts. A comparison of some of these properties is shown in Table 9.4, which shows some striking differences between these two types of block copolymers. Thus, with similar blocks of polystyrene and polyisoprene, the star-branched polymers show a much lower viscosity–molecular weight relation, and, surprisingly, a substantially higher tensile strength.

The low viscosity of these branched polymers is, of course, a natural outcome of their highly branched structure. The higher strength is more difficult to explain, but may be due to the morphology of these polymers, as illustrated in Fig. 9.9, which shows a transmission electron microphotograph

TABLE 9.4 Properties of Star-Branched and Linear SIS Block Copolymers[142]

Type	Molecular weight of S block ($\times 10^{-3}$)	Wt% S	Avg. no. branches	$[\eta]^a$	Tensile strength (MPa)	Strain at break (λ)
Star	20	25.2	6	1.30	41.0	11.7
	21	32.3	7	1.33	40.0	10.3
	20	31.2	9	1.35	43.0	11.5
Linear	20	32.4	2	1.30	34.0	11.3
	14	30.0	2	–	34.5	11.5

[a] Intrinsic viscosity in H_4-furan at 25°C (dl g^{-1}).

Fig. 9.9 Electron microphotograph of a "radial" block copolymer from styrene and isoprene.[318] Ultrathin (~ 50 nm) section of cast film, stained with OsO_4. (Reprinted with permission from Ref. 318. Copyright 1975 American Chemical Society.)

of a very thin film of an OsO_4-stained star-branched styrene–isoprene block copolymer. The remarkable regularity and sharp boundaries of the polystyrene domains in this case may indicate a very high degree of phase separation, leading to the formation of very "pure" polystyrene domains which might be capable of withstanding higher stresses.

d. Other Anionic ABA Triblock Copolymers

Thermoplastic elastomers can also be obtained by using monomer pairs other than styrene and dienes, provided the ABA triblock copolymers have an elastomeric center block and glassy (or crystalline) end blocks. Thus propylene sulfide has been substituted[142, 266] for isoprene for the center block, using α-methylstyrene for the end blocks. This required the formation of the diblock AB, followed by a linking reaction with a dichlorosilane, since the lithium thiolate end groups are incapable of initiating the polymerization of vinyl monomers. The resulting triblock copolymers showed thermoplastic elastomer behavior, as expected.

The anionic ring-opening polymerization of cyclic sulfides was also used[37, 142] to make ABA triblock copolymers having polyisoprene center blocks and polythiabutane end blocks. This required the use of a dilithium initiator to prepare a dilithiopolyisoprene, which then acted as the initiator for the polymerization of thiacyclobutane. In this case the end blocks were not glassy but crystalline ($T_m = 55°C$), rendering the polymer insoluble at room temperature. A similar triblock copolymer with crystalline end blocks ($T_m \sim 200°C$) was prepared from ethylene sulfide.

In another variation on thermoplastic elastomers, the end blocks consisted of polystyrene or poly-α-methylstyrene with a polydimethylsiloxane center block.[142, 242, 248] These could be prepared[248] by sequential polymerization of styrene (or α-methylstyrene) and hexamethylcyclotrisiloxane, using *sec*-butyllithium as initiator, followed by coupling of these diblocks by dimethyldichlorosilane. The resulting triblock copolymers exhibited the expected thermoplastic elastomer behavior.[248]

The anionic synthesis of ABA triblock copolymers with crystalline end blocks has been of particular interest in more recent years. This is because such thermoplastic elastomers could have two obvious advantages over the styrene–diene types, i.e., (i) the crystalline end blocks should render the polymer solvent resistant at moderate temperatures, and (ii) the polymer should offer less resistance to flow, since the two types of blocks need not be strongly incompatible, "phase separation" occurring by crystallization rather than by incompatibility. (It has been noted that the viscosity of these heterophase materials increases as the solubility parameters of the two components diverge, since the flow process really involves the heat of mixing.)

In this connection two recent approaches are worthy of note. One involved the synthesis of a triblock copolymer with a polyisoprene center block and polyester end blocks, prepared by anionic ring-opening polymerization of a lactone (pivalolactone).[95] The other approach[321,322] was to synthesize an ABA triblock copolymer where the B block was either polyisoprene or a mixed structure (1,4- and 1,2-polybutadiene) and the A blocks were high-1,4-polybutadiene, and then to *hydrogenate* the polymer so that the end blocks simulate the structure of polyethylene. Such materials also exhibited thermoplastic elastomer behavior. As expected, the block copolymers with polyester end blocks[95] benefited from the higher melting point of such end blocks, and therefore retained their integrity to higher temperatures.

References

1. M. Morton, *in* "Block Polymers" (S. L. Aggarwal, ed.), p. 1. Plenum, New York, 1970.
2. M. Morton and L. J. Fetters, *in* "High Polymers," Vol. 29 (C. E. Schildknecht and I. Skeist, eds.), p. 286. Wiley, New York, 1977.
3. L. J. Fetters, *J. Res. Natl. Bur. Stand., Ser. A* **70**, 421 (1966).
4. K. Novo, H. Kawazura, T. Moriyama, and S. Yoshioka, *Makromol. Chem.* **83**, 35 (1965).
5. W. E. Smith, F. R. Galiano, D. Rankin, and G. J. Mantell, *J. Appl. Polym. Sci.* **10**, 1659 (1966).
6. Chemical Investors S. A. Netherlands Patent Appl. 6,411,511 (1965); *CA* **63**, 10089d.
7. E. V. Kirkland and W. J. Roberts, U. S. Patent 3,219,725 (1965); *CA* **62**, 10547f.
8. G. J. Mantell, W. E. Smith, F. R. Galiano, and D. Rankin, U. S. Patent 3,732,333 (1973); *CA* **79**, 67066a.
9. E. V. Kirkland and W. J. Roberts, Belgian Patent 636,370 (1963); *CA* **62**, 6590f.
10. J. C. Galin, *Makromol. Chem.* **124**, 118 (1969).
11. J. Furukawa, T. Saegusa, and N. Mise, *Makromol. Chem.* **38**, 244 (1960).
12. J. C. Galin, *Peint. Pigm., Vernis* **43**, 531 (1967); *CA* **67**, 117620b.
13. N. S. Nametkin, S. G. Durgar'yan, I. N. Kozhukhova, and V. N. Filippova, *Vysokomol. Soedin., Ser. A* **16**, 153 (1974); *Polym. Sci. USSR (Engl. Transl.)* **16**, 182 (1974).
14. N. S. Nametkin, S. G. Durgar'yan, I. N. Kozhukhova, and V. N. Filippova, *Vysokomol. Soedin., Ser. A* **17**, 1206 (1975); *Polym. Sci. USSR (Engl. Transl.)* **17**, 1381 (1975).
15. Asahi Chemical Co. Ltd., Japanese Patent 70/20,300 (1970).
16. C. King and F. T. Wallenberger, U. S. Patent 3,538,195 (1970); *CA* **74**, 32623z.
17. Y. Yamashita, Y. Nakamura, and S. Kojima, *J. Polym. Sci.* **11**, 823 (1973).
18. Y. Minoura, M. Mitoh, A. Tabuse, and Y. Yamada, *J. Polym. Sci., Polym. Chem. Ed.* **7**, 2753 (1969).
19. W. P. Baker, Jr., French Patent 1,357,933 (1964); *CA* **62**, 4184h.
20. W. P. Baker, Jr., U. S. Patent 3,274,295 (1966); *CA* **66**, 19138k.
21. J. J. O'Malley, Ger. Offen 2,136,721 (1972); *CA* **77**, 6086u.
22. T. F. Niemann, U. S. Patent 3,700,756 (1972); *CA* **78**, 30541e.
23. H. Takeda and K. Noro, *Kobunshi Kagaku* **21**, 445, 452, 459 (1964); *CA* **62**, 10537e.
24. W. P. Baker, Jr., French Patent 1,357,936 (1964); *CA* **62**, 1817b.
25. Phillips Petroleum Co., British Patent 895,980 (1962); *CA* **58**, 14140a.
26. H. E. De la Mare and F. E. Neumann, Ger. Offen. 1,938,289 (1970); *CA* **72**, 90999w.
27. J. C. Falk and R. J. Schlott, *Macromolecules* **4**, 152 (1971).

28. W. L. Hergenrother and R. J. Ambrose, *J. Polym. Sci., Polym. Chem. Ed.* **12**, 2613 (1974).
29. Firestone Tire & Rubber Co., Netherlands Patent Appl. 7,306,814 (1974); *CA* **83**, 29541b.
30. H. L. Hsieh, F. X. Mueller, Jr., and W. R. Busler, Ger. Offen. 1,805,864 (1970).
31. F. X. Mueller, Jr. and H. L. Hsieh, U. S. Patent 3,585,257 (1971); *CA* **73**, 78331k.
32. J. P. Billot, A. Douy, and B. Gallot, *Makromol. Chem.* **177**, 1889 (1976).
33. B. Perly, A. Douy, and B. Gallot, *Makromol. Chem.* **177**, 2569 (1976).
34. J. P. Billot, A. Douy, and B. Gallot, *Makromol. Chem.* **178**, 1641 (1977).
35. R. P. Foss, Ger. Offen. 2,307,740 (1973); *CA* **80**, 97171u.
36. J. G. Balas, U. S. Patent 3,448,176 (1969); *CA* **71**, 51058y.
37. M. Morton and S. L. Mikesell, *Polym. Prepr., Am. Chem. Soc., Div. Polym. Chem.* **13** (1), 61 (1972); *J. Macromol. Sci., Chem.* **A7**, 1391 (1973).
38. W. Cooper, P. T. Hale, and J. S. Walker, *Polymer* **15**, 175 (1974).
39. J. Furukawa, S. Yamashita, N. Harada, and Y. Arai, Japanese Patent 72/08,641 (1972); *CA* **77**, 63094u.
40. A. Gourdenne, *in* "Block Polymers" (S. L. Aggarwal, ed.), p. 277. Plenum, New York, 1970.
41. A. Gourdenne and P. Sigwalt, *Eur. Polym. J.* **3**, 481 (1967); *CA* **67**, 91142v.
42. A. Gourdenne and P. Sigwalt, *Bull. Soc. Chim. Fr.* **7**, 2249 (1967); *CA* **67**, 91147.
43. A. Gourdenne and P. Sigwalt, *Bull. Soc. Chim. Fr.* **10**, 3685 (1967); *CA* **68**, 13446d.
44. W. G. Davies and D. P. Jones, British Patent 1,376,446 (1974); *CA* **83**, 11146s.
45. Shell Internationale Research Maatschappij, N.V., Netherlands Patent Appl. 6,514,631 (1966); *CA* **66**, 11635z.
46. N. F. Kovalev, A. A. Korotkov, G. N. Petrov, V. N. Reikh, G. F. Lisochkin, L. V. Dugina, and L. A. Eventova, *Kauch. Rezina* **25**, 2 (1966); *CA* **66**, 95949r.
47. A. Gourdenne and P. Sigwalt, *Bull. Soc. Chim. Fr.* **10**, 3678 (1967); *CA* **68**, 13445c.
48. J. F. Henderson and J. Darcy, French Patent 1,522,823 (1968); *CA* **70**, 116065z.
49. A. W. Shaw, Ger. Offen. 2,045,622 (1971); *CA* **75**, 7117a.
50. E. Mueller, R. Mayer-Mader, and K. Dingers, *Nuova Chim.* **49**, 66 (1973); *CA* **79**, 5661g.
51. P. Radici, G. Bianchi, D. Colombo, and P. Colombo, Ger. Offen. 2,557,360 (1976); *CA* **85**, 143787c.
52. Shell Internationale Research Maatschappij N.V., Netherlands Patent Appl. 6,912,915 (1969); *CA* **72**, 79672p).
53. F. Kobayashi, T. Iwasaki, K. Tomita, and N. Utsumi, Japanese Patent 72/25,879 (1972); *CA* **78**, 16777e.
54. B. H. Werner and R. A. Hayes, U. S. Patent 3,758,631 (1973); *CA* **80**, 28175m.
55. P. R. Schaeffer and N. E. Steely, U. S. Patent 3,511,893 (1970); *CA* **73**, 4740w.
56. M. Matzner, A. Noshay, and J. E. McGrath, U. S. Patent 3,770,849 (1973).
57. J. Lincoln, British Patent 1,150,725 (1969); *CA* **71**, 3910y.
58. M. Matzner and J. E. McGrath, U. S. Patent 3,657,385 (1972); *CA* **77**, 35521n.
59. J. E. McGrath, L. M. Robeson, and M. Matzner, *Polym. Prepr., Am. Chem. Soc., Div. Polym. Chem.* **14**(2), 1032 (1973).
60. S. Gaskin and E. Nield, Ger. Offen. 2,210,348 (1972); *CA* **77**, 165536g.
61. A. E. Allegrezza, Jr., R. W. Lenz, J. Cornibert, and R. H. Marchessault, *Polym. Prepr., Am. Chem. Soc., Div. Polym. Chem.* **14**(2), 1232 (1973).
62. R. T. Ingwall, H. A. Scherage, N. Lotan, A. Berger, and E. Katchalski, *Biopolymers* **6**, 331 (1968).
63. J. S. Spriggs, U. S. Patent 2,828,345 (1958).
64. L. G. Lunsted, British Patent 722,746 (1955).
65. D. R. Jackson and L. G. Lundsted, U. S. Patent 2,677,700 (1954).
66. D. R. Jackson and L. G. Lundsted, U. S. Patent 3,036,130 (1962).
67. A. S. Teot, Canadian Patent 698,568 (1964).
68. R. Perret and A. Skoulios, *C. R. Hebd. Seances Acad. Sci., Ser. C* **268**, 230 (1969); *CA* **70**, 78632x.

69. R. Perret and A. Skoulios, *Makromol. Chem.* **156**, 143 (1972).
70. J. V. Koleske, R. M. J. Roberts, and F. P. Del Giudice, U. S. Patent 3,670,045 (1972).
71. F. E. Bailey, Jr. and H. G. France, U. S. Patent 3,312,753 (1967); *CA* **67**, 3393z.
72. Y. Ogawa, K. Notani, Y. Yamakawa, and N. Awata, Japanese Patent 72/28,719 (1972); *CA* **78**, 98897f.
73. S. Boileau and P. Sigwalt, *Makromol. Chem.* **171**, 11 (1973).
74. F. E. Bailey, Jr. and H. G. France, U. S. Patent 3,029,216 (1962); *CA* **57**, 2440c.
75. A. E. Skoulios, G. Tsouladze, and E. Franta, *J. Polym. Sci., Part C* **4**, 507 (1963).
76. D. R. Jackson and L. G. Lundsted, U. S. Patent 3,036,118 (1962).
77. L. G. Lundsted, U. S. Patent 2,674,619 (1954).
78. T. Vaughn, D. R. Jackson, and L. G. Lundsted, *J. Am. Oil Chem. Soc.* **29**, 240 (1952).
79. I. R. Schmolka, L. R. Bacon, *J. Am. Oil Chem. Soc.* **44**, 559 (1967); *CA* **67**, 118450h.
80. General Motors Corp., Canadian Patent 939,868 (1970).
81. E. T. Cline, German Patent 1,139,974 (1962); *CA* **58**, 5803c.
82. E. I. du Pont de Nemours & Co., British Patent 807,589 (1959); *CA* **55**, 2199c.
83. R. H. Gobran and S. W. Osborn, U. S. Patent 3,504,050 (1970); *CA* **72**, 122651v.
84. R. H. Gobran and S. W. Osborn, French Patent 1,524,572 (1968); *CA* **71**, 13554z.
85. D. A. MacKillop, *J. Polym. Sci., Part B* **8**, 199 (1970).
86. S. Boileau and P. Sigwalt, *C. R. Hebd. Seances Acad. Sci.* **261**, 132 (1965); *CA* **63**, 13421b.
87. S. Boileau and P. Sigwalt, *Makromol. Chem.* **131**, 7 (1970); *CA* **72**, 91102k.
88. E. E. Bostick, U. S. Patent 3,337,497 (1967); *CA* **67**, 82684w.
89. E. E. Bostick and W. A. Fessler, Ger. Offen. 2,049,547 (1971); *CA* **75**, 21478a.
90. E. E. Bostick and J. J. Zdaniewski, Ger. Offen. 2,048,914 (1971); *CA* **75**, 21496e.
91. E. E. Bostick, *Polym. Prepr., Am. Chem. Soc., Div. Polym. Chem.* **10**(2), 877 (1969).
92. Farbenfabriken Bayer AG, French Patent 1,598,865 (1970); *CA* **74**, 77234q.
93. T. Tabuchi, K. Nobutoki, and H. Sumitomo, *Kogyo Kagaku Zasshi* **71**, 1926 (1968); *CA* **70**, 58488b.
94. W. H. Sharkey, Ger. Offen. 2,013,553 (1970); *CA* **74**, 13980x.
95. R. P. Foss, H. W. Jacobson, H. N. Cripps, and W. H. Sharkey, *Macromolecules* **9**, 373 (1976).
96. E. Franta and P. Rempp, *C. R. Hebd. Seances Acad. Sci.* **254**, 674 (1962).
97. Dunlop Co. Ltd., French Patent 1,566,887 (1969); *CA* **72**, 4219n.
98. P. T. Hale and G. A. Pope, *Eur. Polym. J.* **11**, 677 (1975).
99. D. E. Eaves, J. S. Walker, and A. Stokes, Ger. Offen. 2,025,051 (1970); *CA* **74**, 43305y.
100. R. A. Godfrey and G. W. Miller, *J. Polym. Sci., Polym. Chem. Ed.* **7**, 2387 (1969).
101. A. Marsiat and Y. Gallot, *Makromol. Chem.* **177**, 241 (1976).
102. M. Morton, A. Rembaum, and E. E. Bostick, *J. Appl. Polym. Sci.* **8**, 2707 (1964).
103. M. Morton and A. Rembaum, U. S. Patent 3,051,684; *CA* **57**, 16890.
104. Esso Research and Engineering Co., British Patent 884,974 (1961); *CA* **58**, 11482g.
105. R. K. Graham, J. R. Panchak, and M. J. Kampf, *J. Polym. Sci.* **44**, 411 (1960).
106. G. Champetier, M. Fontanille, and P. Sigwalt, *C. R. Hebd. Seances Acad. Sci.* **250**, 3653 (1960).
107. T. F. Niemann, U. S. Patent 3,709,101 (1971); *CA* **76**, 25836f.
108. T. Yamaguchi and T. Goto, Ger. Offen. 1,934,593 (1970); *CA* **72**, 79651f.
109. T. Yamaguchi and T. Goto, Japanese Patent 71/17,127 (1971); *CA* **75**, 152332w.
110. T. F. Niemann, U. S. Patent 3,609,100 (1971); *CA* **76**, 25837k.
111. T. F. Niemann, U. S. Patent 3,699,191 (1972); *CA* **78**, 59020v.
112. H. Ailhand, Y. Gallot, and A. Skoulios, *Makromol. Chem.* **140**, 179 (1970).
113. G. F. D'Alelio and R. C. Evers, *J. Polym. Sci., Polym. Chem. Ed.* **5**, 813 (1967).
114. R. S. Nevin and E. M. Pearce, *J. Polym. Sci., Part B* **3**, 487 (1965).
115. S. Fukawa, K. Satake, T. Yamada, and K. Sakamoto, Japanese Patent 75/141,691 (1975); *CA* **84**, 136297u.

116. H. A. J. Schepers and B. C. Roest, Ger. Offen. 2,231,993 (1973); *CA* **78**, 98283c.
117. J. M. Guyon-Gellin, J. Gole, and J. P. Pascault, *J. Appl. Polym. Sci.* **19**, 3173 (1975).
118. P. C. Juliano, Ger. Offen. 2,164,469 (1972); *CA* **77**, 115360z.
119. P. C. Juliano, U. S. Patent 3,663,650 (1972); *CA* **77**, 127417m.
120. P. K. Seow, Y. Gallot, and A. Skoulios, *Makromol. Chem.* **176**, 3153 (1975).
121. J. H. Carter and F. W. Michelotti, *Polym. Prepr., Am. Chem. Soc., Div. Polym. Chem.* **5**(2), 614 (1964).
122. P. Claes and G. Smets, *Makromol. Chem.* **44-46**, 212 (1961).
123. A. E. Barnabeo, W. S. Creasy, and L. M. Robeson, *J. Polym. Sci., Polym. Chem. Ed. A-1* **13**, 1979 (1975).
124. Shell International Research Maatschappij N.V., Netherlands Patent Appl. 6,513,888 (1966); *CA* **65**, 10691c.
125. S. Enomoto and H. Wada, *Kobunshi Kagaku* **26**, 673 (1969); *CA* **72**, 67307k.
126. H. E. De la Mare and C. H. Wilcoxen, Jr., Ger. Offen. 1,940,278 (1970); *CA* **72**, 122643u.
127. W. R. Haefele, A. W. Shaw, and E. T. Bishop, British Patent 1,156,300 (1969); *CA* **71**, 71725x.
128. H. L. Hsieh and W. J. Trepka, U. S. Patent 3,410,836 (1968); *CA* **70**, 29568x.
129. R. P. Zelinski, U. S. Patent 3,287,333 (1966).
130. Phillips Petroleum Co., British Patent 964,478 (1964); *CA* **61**, 10853g.
131. R. P. Zelinski and H. L. Hsieh, U. S. Patent 3,078,254 (1963); *CA* **58**, 10324d.
132. R. P. Zelinski, U. S. Patent 3,251,905 (1966); *CA* **65**, 2370h.
133. E. Shiratsuchi, S. Hayashi, and C. Nozaki, Japanese Patent 70/01,629 (1970); *CA* **72**, 101624y.
134. R. E. Cunningham, M. Auerbach, and W. J. Floyd, *J. Appl. Polym. Sci.* **16**, 163 (1972).
135. Y. K. Wei, Canadian Patent 786,865 (1968); *CA* **69**, 44458j.
136. Shell International Research Maatschappij N.V., Netherlands Patent Appl. 6,405,416 (1964); *CA* **62**, 10648e.
137. Shell Internationale Research Maatschappij N.V., Belgian Patent 671,460 (1966); British Patent 1,014,999 (1966); French Patent 1,418,831 (1965).
138. R. E. Cunningham and M. R. Treiber, *J. Appl. Polym. Sci.* **12**, 23 (1968); *CA* **68**, 87981w.
139. Polymer Corp. French Patent 1,573,989 (1969); *CA* **72**, 56482n.
140. M. Morton, "Encyclopedia of Polymer Science and Technology," Vol. 15, p. 508. Wiley, New York, 1971.
141. M. Morton and S. H. Goh, *Polym. Prepr., Am. Chem. Soc., Div. Polym. Chem.* **14**(1), 92 (1973).
142. M. Morton, *J. Polym. Sci., Polym. Symp.* **60**, 1 (1978).
143. M. Matsuo, T. Ueno, H. Horino, S. Chujyo, and H. Asai, *Polymer* **9**, 425 (1968).
144. M. Morton, J. E. McGrath, and P. C. Juliano, *J. Polym. Sci., Part C* **26**, 99 (1969).
145. M. Morton, L. J. Fetters, F. C. Schwab, C. R. Strauss, and R. F. Kammereck, *Int. Synth. Rubber Symp., Lect., 4th, 1969* p. 70 (1969).
146. M. Morton, *Polym. Prepr., Am. Chem. Soc., Div. Polym. Chem.* **10**, 512 (1969); *in* "Block Polymers" (S. L. Aggarwal, ed.), p. 1. Plenum, New York, 1970.
147. Badische Anilin und Soda-Fabrik AG, French Patent 1,534,150 (1968); *CA* **71**, 13978v.
148. H. E. De la Mare and E. F. Bullard, Ger. Offen. 1,905,422 (1969); *CA* **71**, 92470v.
149. T. Okamura and T. Yoshitake, Japanese Patent 72/05,812 (1972); *CA* **77**, 21198a.
150. I. Ichikawa, T. Miki, and S. Narisawa, Japanese Patent 76/73,092 (1976); *CA* **85**, 124622d.
151. E. B. Bradford and E. Vanzo, *J. Polym. Sci., Polym. Chem. Ed.* **6**, 1661 (1968).
152. R. N. Cooper, Jr., U. S. Patent 3,030,346 (1960); *CA* **57**, 2434a.
153. K. Trukenbrod, G. Weber, *Chim. Ind., Genie Chim.* **104**, 1271 (1971); *CA* **75**, 99133d.
154. Polymer Corp. Ltd., British Patent 1,103,939 (1968); *CA* **68**, 79397v.
155. L. A. Utracki, R. Simha, and L. J. Fetters, *J. Polym. Sci., Polym. Phys. Ed.*, **6**, 2051 (1968).

156. L. A. Utracki and R. Simha, *Polym. Prepr., Am. Chem. Soc., Div. Polym. Chem.* **9**(1), 742 (1968).
157. L. M. Oberlin, U. S. Patent 3,959,412 (1976); *CA* **85**, 79397w.
158. Minnesota Mining and Mfg. Co., Fr. Demande 2,245,746 (1975); *CA* **83**, 180461w.
159. H. L. Hsieh, U. S. Patent 3,402,159 (1968); *CA* **69**, 97570w.
160. R. E. Dollinger, U. S. Patent 3,297,793 (1967); *CA* **66**, 66448w.
161. R. E. Dollinger and R. L. Huxtable, U. S. Patent 3,356,763 (1967); *CA* **68**, 22360.
162. N. F. Keckler, South African Patent 6,705,812 (1968); *CA* **70**, 58765q.
163. Farbenfabriken Bayer AG, Fr. Demande 2,008,521 (1968); *CA* **73**, 46416a.
164. K. Yamaguchi, K. Yuji, T. Ibaragi, K. Toyomoto, and K. Sakamoto, Ger. Offen. 1,963,038 (1970).
165. Farbenfabriken Bayer AG, British Patent 1,218,147 (1971); *CA* **74**, 88383p.
166. H. Sutter and M. Beck, Ger. Offen. 1,939,420 (1971); *CA* **74**, 126440q.
167. J. F. Marsh, British Patent 1,109,616 (1968); *CA* **69**, 3324u.
168. U. S. Patent 3,231,635 (1966); *CA* **64**, 9836f.
169. S. Horiya, S. Asai, and C. Saito, Japanese Patent 73/20,038 (1973); *CA* **80**, 134193a.
170. A. Douy and B. Gallot, *Makromol. Chem.* **156**, 81 (1972).
171. J. Minoux and M. Leng, *C. R. Hebd. Seances Acad. Sci.*, **252**, 277 (1961).
172. Laboratoire de recherche et de controle du Caoutchouc, French Patent 1,209,992 (1960).
173. R. P. Zelinski, Belgian Patent 661,095 (1965); *CA* **65**, 4080f.
174. F. D. Moss and J. F. Mathews, U. S. Patent 3,390,207 (1968); *CA* **69**, 36895j.
175. G. Holden and R. Milkovich, U. S. Patent 3,265,765 (1966).
176. Shell Internationale Research Maatschappij N.V., British Patent 1,035,873 (1966); *CA* **65**, 12389.
177. G. Holden and R. Milkovich, Belgian Patent 627,652 (1973); *CA* **60**, 14714f.
178. Shell Internationale Research Maatschappij N.V., Netherlands Patent Appl. 6,603,846 (1966).
179. M. Szwarc, M. Levy, and R. Milkovich, *J. Am. Chem. Soc.* **78**, 2656 (1956).
180. M. Szwarc, *Nature* (*London*) **178**, 1168 (1965); *CA* **51**, 11815i.
181. S. Schlick and M. Levy, *J. Phys. Chem.* **64**, 883 (1960).
182. A. A. Korotkov, L. A. Shibayev, L. M. Pyrakov, V. G. Aldoshin, and S. Ya. Frenkel, *Vysokomol. Soedin.* **1**, 443 (1959).
183. A. Rembaum, F. R. Ells, R. C. Morrow, and A. V. Tobolsky, *J. Polym. Sci.* **61**, 155 (1962).
184. R. J. Angelo, R. M. Ikeda, and M. L. Wallach, *Polymer* **6**, 141 (1965); *CA* **63**, 3064b.
185. Shell Internationale Research Maatschappij N.V., Netherlands Patent Appl. 6,617,843 (1967); *CA* **68**, 69978w; 6,909,189 (1969); *CA* **72**, 91154d.
186. L. J. Fetters and M. Morton, *Macromolecules* **2**, 453 (1969).
187. H. L. Hassell, Ger. Offen. 2,045,621 (1971); *CA* **74**, 142672.
188. F. D. Moss and A. R. Bean, Jr., U. S. Patent 3,465,065 (1969); *CA* **71**, 9246w.
189. J. Terrisse and F. LeMuer, Fr. Demande 2,138,465 (1973); *CA* **79**, 79494b.
190. D. N. Cramond and J. R. Urwin, *Aust. J. Chem.* **21**, 1835 (1968).
191. N. Corbin and J. Prud'homme, *J. Polym. Sci.* **14**, 1645 (1976).
192. J. Prud'homme and S. Bywater, *Macromolecules* **4**, 543 (1971).
193. T. Inoue, T. Soen, and H. Kawai, *J. Polym. Sci., Part B* **6**, 75 (1968).
194. T. Inoue, M. Moritani, T. Hashimoto, and H. Kawai, *Macromolecules* **4**, 500 (1971).
195. Polymer Corp. Ltd., British Patent 1,121,978 (1968); *CA* **69**, 78286s.
196. C. Price, A. G. Watson, and M. T. Chow, *Polymer* **13**, 333 (1972).
197. B. Gallot, R. Mayer, and C. Sadron, *Rubber Chem. Technol.* **40**, 932 (1967).
198. Y. Yamashita, H. Matsui, and K. Ito, *J. Polym. Sci., Polym. Chem. Ed.* **10**, 3577 (1972).
199. Y. Yamashita, Y. Murase, and K. Ito, *J. Polym. Sci., Polym. Chem. Ed.* **11**, 435 (1973).
200. E. I. duPont de Nemours & Co., French Patent 1,357,935 (1964); *CA* **62**, 1818a.
201. K. Nobutoki and H. Sumitomo, *Bull. Chem. Soc. Jpn.* **40**, 1741 (1967); *CA* **67**, 82452u.

202. F. Kobayashi, T. Iwasaki, N. Utsumi, and N. Asahara, Japanese Patent 68/26,629 (1968); *CA* **70**, 78742h.
203. Y. Yamashita and T. Hane, *J. Polym. Sci., Polym. Chem. Ed.* **11**, 425 (1973); *CA* **78**, 111835f.
204. Y. Yamashita, Japanese Patent 73/00,792 (1973); *CA* **78**, 111,979f.
205. G. F. D'Alelio and T. F. Hoffend, *J. Polym. Sci., Polym. Chem. Ed.* **5**, 323 (1967).
206. H. Ohnuma, T. Kotaka, and H. Inagaki, *Polymer* **10**, 501 (1969).
207. J. R. Urwin and J. M. Stearne, *Makromol. Chem.* **78**, 194 (1964).
208. T. Kotaka, T. Tanaka, and H. Inagaki, *Polym. J.* **3**, 327 (1972).
209. C. W. Brown and I. F. White, *J. Appl. Polym. Sci.* **16**, 2671 (1972).
210. C. G. Overberger and N. Yamamoto, *J. Polym. Sci., Polym. Chem. Ed.* **4**, 3101 (1966).
211. A. Rembaum and M. Szwarc, *J. Polym. Sci.* **22**, 189 (1956).
212. R. K. Graham, D. L. Dunkelberger, and E. S. Cohn, *J. Polym. Sci.*, **42**, 501 (1960).
213. A. Dondos, P. Rempp, and H. Benoit, *Polymer* **13**, 97 (1972).
214. M. Baer, *J. Polym. Sci., Part A* **2**, 417 (1964).
215. D. Freyss, P. Rempp, and H. Benoit, *J. Polym. Sci., Part B* **2**, 217 (1964).
216. D. Freyss, M. Leng, and P. Rempp, *Bull. Soc. Chim. Fr.* **2**, 221 (1964).
217. G. Finaz, Y. Gallot, J. Parrod, and P. Rempp, *J. Polym. Sci.* **58**, 1363 (1962).
218. D. H. Richards and M. Szwarc, *Trans. Faraday Soc.* **55**, 1644 (1959).
219. G. Finaz, P. Rempp, and J. Parrod, *Bull. Soc. Chim. Fr.* p. 262 (1962); *CA* **57**, 1055f.
220. M. Szwarc, U. S. Patent 3,050,511 (1962); *CA* **58**, 1558f.
221. V. N. Zgonnik, L. A. Shibaev, and N. I. Nikolaev, *Kinet. Mech. Polyreactions, Int. Symp. Macromol. Chem., Prepr., 1969* Vol. 4, p. 319 (1969); *CA* **75**, 64363b.
222. J. J. O'Malley and R. H. Marchessault, *Macromol. Synth.* **4**, 35 (1972).
223. J. J. O'Malley, R. C. Crystal, and P. F. Erhardt, *Polym. Prepr., Am. Chem. Soc., Div. Polym. Chem.* **10**(2), 796 (1969).
224. M. Gervais and B. Gallot, *Makromol. Chem.* **171**, 157 (1973).
225. N. Ito and T. Hosogane, Japanese Patent 73/13,705 (1973); *CA* **80**, 37859q.
226. E. Balcerzyk, H. Pstrocki, and G. Wlodarski, *Rocz. Chem.* **44**, 1583 (1970); *CA* **74**, 32225w.
227. H. Pstrocki, Polish Patent 53,771 (1967); *CA* **68**, 87743v.
228. H. Pstrocki, *Rocz. Chem.* **38**, 899 (1964); *CA* **66**, 2882y.
229. E. Balcerzyk, H. Pstrocki, and G. Wlodarski, *J. Appl. Polym. Sci.* **11**, 1179 (1967).
230. E. I. du Pont de Nemours & Co., French Patent 1,357,934 (1964); *CA* **62**, 1821.
231. S. Boileau, G. Champetier, and P. Sigwalt, *Makromol. Chem.* **69**, 180 (1963).
232. P. Sigwalt, *Chim. Ind. (Milan)* **96**, 909 (1966).
233. F. R. Jones, *Eur. Polym. J.* **10**, 249 (1974).
234. J. W. Dean, *J. Polym. Sci., Part B* **8**, 677 (1970).
235. J. C. Saam, D. J. Gordon, and S. Lindsey, *Macromolecules* **3**, 1 (1970).
236. J. C. Saam and F. W. G. Fearon, *Polym. Prepr., Am. Chem. Soc., Div. Polym. Chem.* **11**(2), 455 (1970).
237. J. C. Saam and F. W. G. Fearon, Ger. Offen. 2,142,595 (1972); *CA* **77**, 6090r.
238. General Electric Co., French Patent 2,134,884 (1973); *CA* **79**, 54055r.
239. J. W. Dean, Ger. Offen. 2,116,837 (1971); *CA* **76**, 73475n.
240. J. C. Saam, A. H. Ward, and F. W. G. Fearon, *Polym. Prepr., Am. Chem. Soc., Div. Polym. Chem.* **13**(1), 524 (1972).
241. J. G. Zilliox, J. E. L. Roovers, and S. Bywater, *Macromolecules* **8**, 573 (1975).
242. M. J. Owen and T. C. Kendrick, *Macromolecules* **3**, 458 (1970).
243. G. L. Gaines, Jr. and C. W. Bender, *Macromolecules* **5**, 82 (1972).
244. J. C. Saam, Ger. Offen. 2,011,088 (1970); *CA* **73**, 121222q.
245. W. G. Davies and D. P. Jones, *Ind. Eng. Chem. Prod. Res. Dev.* **10**, 168 (1971); *CA* **75**, 49823h.
246. T. C. Kendrick and M. J. Owen, Ger. Offen. 1,915,789 (1969); *CA* **71**, 125476g.

247. W. G. Davies and D. P. Jones, *Polym. Prepr., Am. Chem. Soc., Div. Polym. Chem.* **11**(2), 447 (1970).
248. M. Morton, Y. Kesten, and L. J. Fetters, *Polym. Prepr., Am. Chem. Soc., Div. Polym. Chem.* **15**(2), 175 (1974); *Appl. Polym. Symp.* **26**, 113 (1975).
249. J. C. Saam and F. W. G. Fearon, French Patent 2,105,983 (1972); *CA* **78**, 16772z.
250. H. J. Kollmeier and G. Rossmy, Ger. Offen. 2,431,394 (1976); *CA* **84**, 106508w.
251. G. Greber, E. Reese, and A. Balciunas, *Farbe Lack* **70**, 249 (1964); *CA* **61**, 5800q.
252. L. M. Robeson, M. Matzner, L. J. Fetters, and J. E. McGrath, *Polym. Prepr., Am. Chem. Soc., Div. Polym. Chem.* **14**(2), 1063 (1973).
253. D. J. Dunn and S. Krause, *J. Polym. Sci., Polym. Lett. Ed.* **12**, 591 (1974).
254. S. Horiie, S. Asai, and Y. Moriya, Japanese Patent 75/133,291 (1975); *CA* **84**, 106330g.
255. D. Rahlwes and R. G. Kirste, *Makromol. Chem.* **178**, 793 (1977).
256. D. R. Hansen and M. Shen, *Macromolecules* **8**, 903 (1975).
257. G. Champetier, M. Fontanille, A. C. Korn, and P. Sigwalt, *J. Polym. Sci.* **58**, 911 (1962).
258. P. Grosius, Y. Gallot, and A. Skoulios, *Makromol. Chem.* **127**, 94 (1969).
259. P. Grosius, Y. Gallot, and A. Skoulios, *Eur. Polym. J.* **6**, 355 (1970).
260. M. Fontanille and P. Sigwalt, *Bull. Soc. Chim. Fr.* **1**, 4095 (1967); *CA* **68**, 69393b.
261. A. Schindler and J. L. Williams, *Polym. Prepr., Am. Chem. Soc., Div. Polym. Chem.* **10**(2), 832 (1969).
262. H. S. Makowski and D. J. Buckley, Ger. Offen. 2,443,875 (1975); *CA* **83**, 44034g.
263. M. Shima, E. Ogawa, and K. Konishi, *Makromol. Chem.* **177**, 241 (1976).
264. G. F. D'Alelio and T. R. Hoffen, *J. Polym. Sci., Polym. Chem. Ed.* **5**, 1245 (1967).
265. T. Kawai, S. Shiozaki, S. Sonoda, H. Hakagawa, T. Matsumoto, and H. Maeda, *Makromol. Chem.* **128**, 252 (1969).
266. M. Morton, R. F. Kammereck, and L. J. Fetters, *Br. Polym. J.* **3**, 120 (1971); *Macromolecules* **4**, 11 (1971).
267. K. S. Dennis, British Patent 1,138,267 (1968); *CA* **70**, 48021e.
268. Dow Chem. Co., French Patent 1,536,112 (1968); *CA* **71**, 13761q.
269. V. P. Shatalov, L. A. Grigorieva, A. E. Kistereva, V. D. Alekhin, A. R. Samotsvetov, and I. J. Kirchevskaya, British Patent 1,409,956 (1975); *CA* **84**, 75025e.
270. H. A. J. Schepers, Ger. Offen. 2,442,849 (1975); *CA* **83**, 11865a.
271. T. Shimomura, H. Nagata, and Y. Murakami, Ger. Offen. 2,224,616 (1972); *CA* **78**, 98759n.
272. V. P. Shatalov, L. A. Grigorieva, and A. E. Kistereva, USSR Patent 254,773 (1968); *CA* **72**, 122639x.
273. F. E. Neumann, Ger. Offen. 2,015,250 (1970); *CA* **74**, 232315.
274. T. Shimomura, H. Nagata, and Y. Murakami, Japanese Patent 73/00,790 (1973); *CA* **78**, 112535v.
275. K. F. Elgert, E. Seiler, G. Puschendort, and H. J. Kantow, *Makromol. Chem.* **165**, 245 (1973); *CA* **78**, 160688n.
276. G. Karoly, *Polym. Prepr., Am. Chem. Soc., Div. Polym. Chem.* **10**(2), 837 (1969).
277. B. C. Roest and H. A. J. Schepers, Netherlands Patent Appl. 7,104,949 (1972); *CA* **78**, 111943q.
278. T. Kotaka, H. Ohnuma, and H. Inagaki, *Polym. Prepr., Am. Chem. Soc., Div. Polym. Chem.* **11**(2), 660 (1970).
279. G. Hsiue, T. Yasukawa, and K. Murakami, *Makromol. Chem.* **139**, 285 (1970).
280. N. Yamazaki, H. Shirakawa, and S. Kambara, *J. Polym. Sci., Part C* **16**, 1685 (1967).
281. C. L. Lee and O. K. Johannson, *J. Polym. Sci., Polym. Chem. Ed.* **4**, 3013 (1966).
282. F. E. Neumann, Ger. Offen. 2,015,249 (1970); *CA* **74**, 13992c.
283. L. D. McKeever and W. J. Cheng, U. S. Patent 3,728,193 (1973); *CA* **78**, 160553w.
284. S. I. Beilin, E. L. Vollershtein, M. P. Teterina, M. N. Shuarts, and B. A. Dolgoplosk, *Vysokomol. Soedin., Ser. B* **11**, 733 (1969); *CA* **72**, 32267v.
285. W. J. Trepka, *J. Polym. Sci., Part B* **8**, 499 (1970).

286. A. Gourdenne, *Makromol. Chem.* **158**, 271 (1972); *CA* **77**, 114943e.
287. A. Gourdenne, *Makromol. Chem.* **158**, 261 (1972); *CA* **77**, 114942d.
288. M. Kamachi, M. Kurihara, and J. K. Stille, *Macromolecules* **5**, 161 (1972).
289. J. W. Dean, U. S. Patent 3,875,254 (1975); *CA* **83**, 148280a.
290. J. W. Dean, U. S. Patent 3,673,272 (1972); *CA* **77**, 89111x.
291. P. Grosius, Y. Gallot, and A. Skoulios, *Makromol. Chem.* **136**, 191 (1970).
292. J. Heller, J. F. Schimscheimer, R. A. Pasternak, C. B. Kingsley, and J. Moacanin, *J. Polym. Sci., Polym. Chem. Ed.* **7**, 73 (1969).
293. N. S. Nametkin, S. G. Durgar'yan, E. V. Solov'ev, and V. M. Piryatinskii, Ger. Offen. 1,917,863 (1969); *CA* **72**, 33016z.
294. N. S. Nametkin, S. G. Durgar'yan, E. V. Solov'ev, and V. M. Piryatinskii, French Patent 2,041,501 (1971); *CA* **75**, 130634u.
295. A. W. Shaw and E. T. Bishop, U. S. Patent 3,634,549 (1972); *CA* **76**, 128453.
296. R. J. Angelo, R. M. Ikeda, and M. L. Wallach, *Polymer* **6**, 141 (1965).
297. R. G. Hayter and E. T. Bishop, U. S. Patent 3,639,523 (1972); *CA* **76**, 142099.
298. W. R. Haefele, A. W. Shaw, and E. T. Bishop, British Patent 1,156,300 (1969); *CA* **71**, 71725s.
299. E. E. Bostick, U. S. Patent 3,337,497 (1967); *CA* **67**, 82684w.
300. E. E. Bostick, *Polym. Prepr., Am. Chem. Soc., Div. Polym. Chem.* **10**(2), 877 (1969).
301. K. Kamatsu, K. Harata, N. Ohshima, and M. Takahashi, Japanese Patent 75/15,028 (1975); *CA* **83**, 180257j.
302. C. M. Selman, U. S. Patent 3,887,643 (1975); *CA* **83**, 133043b.
303. H. A. J. Schepers and B. C. Roest, Ger. Offen. 2,059,372 (1971); *CA* **75**, 77945.
304. P. T. Hale, D. A. Bird, and G. A. Pop, Ger. Offen. 1,911,241 (1969); *CA* **72**, 4201a.
305. W. Cooper, P. T. Hale, and J. S. Walker, *Polymer* **15**, 175 (1974).
306. E. E. Bostick, U. S. Patent 3,483,270 (1969); *CA* **72**, 44357n).
307. H. Hendus, K. H. Ilers, and E. Ropte, *Kolloid Z. Z. Polym.* **216-217**, 110 (1967).
308. J. F. Beecher, L. Marker, R. D. Bradford, and S. L. Aggarwal, *J. Polym. Sci., Part C* **26**, 117 (1969).
309. M. Matsuo, *Jpn. Plast.* **2**, 6 (1968).
310. P. R. Lewis and C. Price, *Nature (London)* **223**, 494 (1969).
311. D. McIntyre and E. Campos-Lopez, *Macromolecules* **3**, 322 (1970).
312. D. J. Meier, *J. Polym. Sci., Part C* **26**, 81 (1969); *Polym. Prepr., Am. Chem. Soc., Div. Polym. Chem.* **15**, 171 (1979); E. Helfand and Z. Wasserman, *Macromolecules* **9**, 878 (1976); *Polym. Prepr., Am. Chem. Soc., Div. Polym. Chem.* **18**, 290 (1977); A. Keller, E. Pedemonte, and F. M. Willmouth, *Nature (London)* **225**, 538 (1970); *Kolloid Z. Z. Polym.* **238**, 25 (1970); J. Dugloz, A. Keller, and E. Pedemonte, *ibid.* **242**, 1125 (1970).
313. G. Holden, E. T. Bishop, and N. R. Legge, *J. Polym. Sci., Part C* **26**, 81 (1969).
314. P. A. Small, *J. Appl. Chem.* **3**, 71 (1953).
315. R. F. Fedors, *J. Polym. Sci., Part C* **26**, 189 (1969).
316. D. J. Meier, *J. Phys. Chem.* **71**, 1861 (1967); *J. Polym. Sci., Part C* **26**, 81 (1969).
317. T. L. Smith and R. A. Dickie, *J. Polym. Sci., Part C* **26**, 163 (1969).
318. L. K. Bi and L. J. Fetters, *Macromolecules* **8**, 90 (1975); **9**, 732 (1976).
319. H. L. Hsieh, *Rubber Chem. Technol.* **49**, 1305 (1976).
320. B. J. Bauer and L. J. Fetters, *Rubber Chem. Technol.* **51**, 406 (1978).
321. Y. Mohajer, G. L. Wilkes, M. Martin, I. C. Wang, and J. E. McGrath, *Polym. Prepr., Am. Chem. Soc., Div. Polym. Chem.* **21**, 43 (1980).
322. M. Morton, N. C. Lee, and E. R. Terrill, *Polym. Prepr., Am. Chem. Soc., Div. Polym. Chem.* **22**, 136 (1981).

Chapter

10 Synthesis of Branched Polymers

The nonterminating character of anionic polymerization also makes it particularly suitable for the synthesis of branched macromolecules, some of which can be prepared with a high degree of precision not possible by other means. These branched polymers are of two types: (i) graft polymers, consisting of a "main" chain to which are attached branches of the same or different chain unit structure, and (ii) "star"-branched polymers, consisting of a number of branches connected at a central point. Such structures have, of course, been prepared by other mechanisms of polymerization, e.g., free radical and cationic, but the anionic nonterminating type offers distinct advantages in the precision of the synthesis.

1. Graft Copolymers and Comb-Branched Polymers

A distinction is made here between a graft *copolymer*, where branches of a given chemical structure are "grafted" onto a main chain backbone having a *different* chemical structure, and a "comb-branched" polymer, where the branches have the *same* chemical structure as the backbone. The method of synthesis may, of course, be similar for either of these types, and can generally fall into one of two classes: (i) growth of a branch from an active site on a polymer chain, and (ii) coupling of an already-formed polymer chain with a

TABLE 10.1 Anionic Graft Copolymers[a]

Backbone chain	Graft monomer	Initiator system	Solvent	References
Polystyrene	Acrylonitrile	n-C_4H_9Li	Toluene	1
	Styrene	Cumylpotassium	H_4-furan	2, 3
	Isoprene	s-C_4H_9Li	Benzene	4
Chlormethylated polystyrene	Isoprene	n-C_4H_9Li	Benzene/TMEDA[b]	5
	Styrene	Cumylpotassium	H_4-furan	6
	Styrene	n-C_4H_9Li	Benzene/H_4-furan	6
	Methyl methacrylate	n-C_4H_9Li	Toluene, etc.	7
Poly-*p*-chlorostyrene (or copolymers with styrene)	Styrene	Na naphthalene	H_4-furan	8–10
	Acrylonitrile	Na naphthalene	H_4-furan	8–10
	Methyl methacrylate	Na naphthalene	H_4-furan	8–10
	Vinylpyridine	Na naphthalene	H_4-furan	8–10
Poly(styrene-*co*-*p*-bromostyrene)	Styrene	Li naphthalene	H_4-furan	11
	Styrene	Na naphthalene	H_4-furan	12
	Methyl methacrylate	Na naphthalene	H_4-furan	11
Poly(styrene-*co*-methyl methacrylate)	Styrene	Cumylpotassium	H_4-furan	11
Poly(styrene-*co*-acrylonitrile)	Acrylonitrile	Na	H_4-furan	13
	Acrylonitrile	Dimethylsilylmethyl-magnesium chloride	H_4-furan	13a
	Methyl methacrylate	Dimethylsilylmethyl-magnesium chloride	H_4-furan	13a
	4-Vinylpyridine	Dimethylsilylmethyl-magnesium chloride	H_4-furan	13a
	4-Vinylpyridine	Na	H_4-furan	13
Poly(styrene-*co*-vinyl acetate)	Styrene	s-C_4H_9Li	Benzene	14
Poly(*p*-vinylstyrene oxide) (and its styrene copolymers)	Styrene	Cumylpotassium	H_4-furan	15
	Styrene	n-C_4H_9Li	Benzene/H_4-furan	15
Poly(methyl methacrylate)	Polystyrylpotassium[c]	–	H_4-furan	16, 17
	Polystyryllithium	–	Toluene	18

	K alkoxide of poly(ethylene oxide)[c]	–	Toluene or toluene/H_4-furan	19
Polybutadiene	Styrene	n-C_4H_9Li/TMEDA	n-Heptane	20
	Styrene	n-C_4H_9Li/TMEDA	Cyclohexane	21
	α-Methylstyrene	n-C_4H_9Li/TMEDA	Cyclohexane	21
Polybutadiene (halogenated)	Styrene	s-C_4H_9Li	Cyclohexane	22
Poly(butadiene-*co*-2-vinylnaphthalene)	2-Vinylnaphthalene	n-C_4H_9Li	Cyclohexane/H_4-furan	23
Polyisoprene	Styrene	n-C_4H_9Li/TMEDA	n-Heptane	24
	Acrylonitrile	n-C_4H_9Li/TMEDA	n-Heptane	24
	Methyl methacrylate	n-C_4H_9Li/TMEDA	n-Heptane	24
	Styrene	s-C_4H_9Li/TMEDA	Cyclohexane	21, 25
Poly(vinyl chloride)	Styrene	Cumylpotassium	H_4-furan	17
	Styrene	n-C_4H_9Li	H_4-furan	26, 27
	Styrene	Na naphthalene	H_4-furan	28
	Butadiene	Na naphthalene	H_4-furan	29
Polyisobutene (halogenated)	Styrene	s-C_4H_9Li	Cyclohexane	22
Butyl rubber (chlorinated)	Styrene	Na naphthalene	H_4-furan	28
EPDM	Styrene	s-C_4H_9Li/TMEDA	Cyclohexane	30
EPDM (halogenated)	Styrene	s-C_4H_9Li	Cyclohexane	22
EPDM (chlorinated)	Styrene	Na naphthalene	H_4-furan	28
	Isoprene	n-C_4H_9Li	Benzene	28
Polychloroprene	Styrene	Na naphthalene	H_4-furan	28
Polyepichlorohydrin	Styrene	Na naphthalene	H_4-furan	28
Poly(epichlorohydrin-co-ethylene oxide)	Styrene	Na naphthalene	H_4-furan	28
Poly(vinyl alcohol)	Acrylonitrile	Na, K	Liq. NH_3 or inert solvents	31
	Methacrylonitrile	Na, K	Liq. NH_3 or inert solvents	31
	Methyl methacrylate	Na, K	Liq. NH_3 or inert solvents	31
	Formaldehyde	K naphthalene	$(CH_3)_2SO$	32
Poly(2-vinylpyridine)	Styrene	n-C_4H_9Li	Benzene/H_4-furan	33

(*table continues*)

TABLE 10.1 Graft Copolymers (*continued*)

Backbone chain	Graft monomer	Initiator system	Solvent	References
Poly(vinylbenzophenone)	Styrene	Na	H_4-furan	34
	Acrylonitrile	Na	H_4-furan	34
	Methyl methacrylate	Na	H_4-furan	34
	4-Vinylpyridine	Na	H_4-furan	34
Poly(2-vinylfluorene)	Methyl methacrylate	Li naphthalene	H_4-furan	35
Poly(vinylnaphthalene)	Styrene	Na	H_4-furan	36
	Ethylene oxide	Na	None	36
Poly(benzylidene-4-vinylaniline)	Acrylonitrile, methyl methacrylate	K	H_4-furan	37
Poly(cinnamylidene-4-vinylaniline)	Acrylonitrile, methyl methacrylate	K	H_4-furan	37
Poly(3,3-diphenyl-1-propene)	Styrene, MMA, acrylonitrile, vinylpyridine	Na or Li naphthalene	H_4-furan	38, 39
Poly(2,6-dimethyl-1,4-phenylene ether)	Styrene, isoprene	n-C_4H_9Li/TMEDA	H_4-furan or toluene/H_4-furan	40
	MMA, hexamethylcyclotrisiloxane, phenyl isocyanate	n-C_4H_9Li/TMEDA	H_4-furan or toluene/H_4-furan	40
Starch or dextrin	Methyl methacrylate	K naphthalene	$(CH_3)_2SO$	41
	Acrylonitrile	K naphthalene	$(CH_3)_2SO$	42
	Methacrylonitrile	Li, Na, or K naphthalene	$(CH_3)_2SO$	43
	4-Vinylpyridine	K naphthalene	$(CH_3)_2SO$	44
	Ethylene oxide	K naphthalene	$(CH_3)_2SO$	45
	Formaldehyde	K naphthalene	$(CH_3)_2SO$	32
	N-Carboxy anhydrides	K naphthalene	$(CH_3)_2SO$	46
Cellulose	MMA, acrylonitrile, methacrylonitrile	Na or K	Liq. NH_3 or other inert solvents	31
	Benzyl methacrylate	Na or K	Liq. NH_3	47
Cellulose acetate	Acrylonitrile	Li, Na, or K naphthalene	H_4-furan, DMF, or $(CH_3)_2SO$	48

	Methacrylonitrile	Li, Na, or K naphthalene	H_4-furan, DMF, or $(CH_3)_2SO$	48
	N-Carboxy anhydrides	Li, Na, or K naphthalene	Dioxane or $(CH_3)_2SO$	49
Wood polysaccharide 4-*O*-Methyl-glucuronxylan	Styrene	Cumylpotassium	H_4-furan	50
Polyisoprene	Pivalolactone	Three-stage reaction		51
Poly(isobutene-co-4-methylstyrene)	Pivalolactone	1. C_4H_9Li/TMEDA		52
Poly(isobutene-co-vinyltoluene)	Pivalolactone	2. $CO_2 + H^+$		52
EPDM	Pivalolactone	3. $(n\text{-}C_4H_9)_4NOH$		53
Poly(ethyl acrylate)	Pivalolactone	$(n\text{-}C_4H_9)_4NOH$	H_4-furan	54
Poly(chloroprene-co-methacrylic acid)	Pivalolactone	$(n\text{-}C_4H_9)_4NOH$	H_4-furan	54
Poly(ethylene-alt-ethyl acrylate)	Pivalolactone	$(n\text{-}C_4H_9)_4NOH$	H_4-furan	54

[a] Includes comb-branched homopolymers.
[b] Tetramethylethylenediamine.
[c] Polymer linking reaction.

functional group located on the backbone polymer chain. Again it is clear that both of these methods are also available to other mechanisms of polymerization, but the anionic nonterminating systems offer the unique advantage that the branches, whether grown out from the main chain or coupled to the main chain, can be of remarkable uniformity in length.

It should be remembered, however, that even in an anionic system there is generally no control on the placement of the active sites along the main backbone chain, so that, although the *length* of the branches may be near monodisperse, the *distribution* of the branch sites would generally be random.

Table 10.1 contains a listing of graft copolymers (and comb-branched polymers) described in the literature. Unless otherwise stated, the grafting was done by first treating the backbone polymer with the indicated "initiator," after which the monomer was polymerized by initiation at the active site located on the backbone polymer chain. As might be expected, most of these active sites in the backbone polymer were created by reaction with an alkali metal or organometal which was capable of displacing a hydrogen from a hydrocarbon polymer, or some other reactive atom from other polymers. Hence the growth of the graft branch invariably was initiated by an alkali metal atom or similar species and resulted in a nonterminating anionic polymerization.

It can be seen that the vast majority of cases involve graft copolymerization, so that the branches have a different chemical composition than the main chain, and this is not surprising, considering the objectives involved in the synthesis of such polymers. It is known that branched polymers behave differently from linear chains, both in their processing characteristics (rheology) and their mechanical properties. However, the really interesting differences occur when the branched polymer is a graft copolymer, since the branches are generally *incompatible* with the backbone chain, resulting in a two-phase system. Such heterophase materials often exhibit specially useful properties. Some of these are discussed in the case of block copolymers (Chapter 9).

2. Star-Branched Polymers

The synthesis of star-branched, or "radial," polymers is accomplished by linking together linear polymers which have a functional group at *one* end. Probably the first attempt to synthesize well-defined polymers of this type was the work of Schaefgen and Flory[55] in 1948. They prepared four-armed and eight-armed star-branched polymers by condensation of ε-aminocaproic acid with a tetrabasic and an octabasic carboxylic acid, respectively. In that case, the terminal functional group was an amino group, which linked up with the polybasic carboxylic acid to yield the branched polymers. It should be noted, however, that the branches in this polymer had the usual "most probable" distribution of chain lengths but that the molecular weight distribution of the branched species corresponded to the relation[55]

$$M_w/M_n = 1 + 1/f \tag{10.1}$$

where f is the number of branches. Hence the M_w/M_n values for the tetra-star and octa-star polymers would be 1.25 and 1.125, respectively.

The nonterminating nature of anionic polymerization, on the other hand, makes it possible to generate terminally functional linear polymers having the narrow Poisson type of molecular weight distribution, which can then be linked together by suitable polyfunctional agents to yield star-branched polymers in which all the "arms" have a remarkable uniformity. The most popular method for the synthesis of this type of polymer has involved homogeneous organolithium polymerization, followed by a linking reaction between the lithium chain end and the linking agent. A recent review[56] has collated the various attempts to prepare such polymers, their characterization, as well as their solution properties.

The first example of the preparation of star-branched polymers by anionic synthesis was the work of Morton, Helminiak, Gadkary, and Bueche[57] in 1962. They prepared and studied the properties of three-armed and four-armed polystyrene which resulted from the reaction of methyltrichlorosilane and tetrachlorosilane, respectively, with polystyryllithium. The types of polymers prepared in this way have been limited by the relatively few monomers which can yield nonterminating polymerizations, and a number of linking agents, other than the chlorosilanes, have been used.

However, the best results have been obtained with the combination of organolithium polymerizations and linking with chlorosilanes. In this manner, star-branched polymers of excellent purity have been prepared with up to 56 arms! These have provided excellent models for the experimental evaluation of theories about the solution properties and rheological behavior of branched macromolecules.[56]

A comprehensive listing of the various star-branched polymers described in the literature is provided in Table 10.2. Again, the reader should be cautioned that no attempt has been made to examine the validity of the claims made with regard to the success of these syntheses, which should be judged in each separate case. In this connection, some valuable critical evaluation can be found in the recent review quoted above.[56]

It will be noted from Table 10.2 that two basic methods have been used to link together the organometallic chain ends (mainly organolithium) to form these star-branched polymers, i.e., (i) using a polyfunctional linking agent carrying reactive groups capable of "condensing" with the metal atom (Cl, Br, etc.), or (ii) using a di- or polyvinyl compound, e.g., divinylbenzene. There are

TABLE 10.2 Anionic Star-Branched Polymers[a]

Initiator	Solvent	Linking agent	No. of arms	References
		Polystyrene		
n-C_4H_9Li	Benzene	$SiCl_4$	3 and 4	57
	Benzene/H_4-furan	1,2,4-Trichloromethylbenzene	3	58
α-Phenylethylpotassium	H_4-furan	1,2,4,5-Tetrachloromethylbenzene	3 and 4	59
n-C_4H_9Li	Benzene/H_4-furan	1,2,4,5-Tetrachloromethylbenzene	2–11	60
	Benzene/H_4-furan	Hexa-*p*-(chloromethyl)phenylbenzene	6	60a
	Benzene/H_4-furan	1,2-Bis(trichlorosilyl)ethane	3 and 4	61
	Benzene/H_4-furan	Cyclic trimer of phosphonitrilic chloride	2, 4, and 6	61
n-C_4H_9Li	Benzene/H_4-furan	*p*- and *m*-divinylbenzenes (DVB)	3.4–22.3	62, 63
Cumylpotassium	Benzene/H_4-furan	*p*- and *m*-divinylbenzenes (DVB)	3.4–22.3	62, 63
s-C_4H_9Li	Benzene or benzene/H_4-furan	*p*- and *m*-divinylbenzenes (DVB)	6–15.5	64, 65
n-C_4H_9Li	Benzene/H_4-furan	$SiCl_4$	3 and 4	66
Cumylpotassium or amylpotassium	H_4-furan or toluene/H_4-furan	2,4,6-Triallyloxytriazine	3	67–69
Cumylpotassium	H_4-furan	1,1,4,4-Tetraphenyl-1,4-bis(diallyloxytriazine)-butane	4	70
s-C_4H_9Li	Benzene	1,2-Bis(dichloromethylsilyl)-ethane	4	71
s-C_4H_9Li	Benzene	1,2-Bis(trichloromethylsilyl)-ethane	6	72
Benzylsodium or cumylpotassium	H_4-furan	1,3,5-Tribromomethylbenzene or 1,3,5-trichloromethylbenzene	3	73
n-C_4H_9Li/DVB	H_4-furan	—	Thousands	74

		Polybutadiene		
n-C_4H_9Li	Cyclohexane	$SiCl_4$	4	75
	Cyclohexane	CH_3SiCl_3	3	75
	Hexane	$SiCl_4$	4	76
	Cyclohexane	$SnCl_4$	4	77, 78
	Benzene	$SiCl_4$	4	79
t-C_4H_9Li	?	Dimethylterephthalate	4	80
		Polyisoprene		
s-C_4H_9Li	n-Heptane/H_4-furan	CH_3SiCl_3, $SiCl_4$	3 and 4	81
	Cyclohexane	CH_3SiCl_3	3	82
	Benzene	$SiCl_4$	3	82
	Benzene	1,2-Bis(dichloromethylsilyl)ethane	4	83
	Benzene	1,2-Bis(trichloromethylsilyl)ethane	6	83
	Benzene	Octa- and dodecachlorosilanes	8–18	84, 85
	Benzene	Divinylbenzene	10–56	86

[a] A tri-star polysiloxane has been synthesized[87] by anionic copolymerization of a mixture of hexamethylcyclotrisiloxane and hexaphenylcyclotrisiloxane by the trifunctional initiator $(LiOCH_2CH_2)_3N$.

advantages and disadvantages in each case. Thus the halogen–metal reaction can be plagued both by steric effects and interconversion reactions. Although the metal–halogen interconversion reaction can be easily avoided by using silicon halides, the latter still may suffer from steric interferences. For example, in the case of silicon tetrachloride, there is complete reaction with polybutadienyllithium, but this is not so for polystyryl- or polyisoprenyllithium, the product being a mixture of three-arm and four-arm species.[57,82] This steric problem can be overcome by using silicon halides in which the silicon atoms are separated by intervening carbon atoms, e.g., 1,2-bis(dichloromethylsilyl)-ethane and similar species, which permit the formation of star-branched polymers with up to 18 arms.[83–85]

In the case of the polyvinyl linking agents such as divinylbenzene, the reaction can also lead to formation of multiarm branched polymers (up to 56 arms[86]), but the stoichiometry is not exact or predictable and there is a distribution of molecular weights of the branched polymers, as might be expected for a mixture of polymers having varying numbers of branches.

References

1. M. B. Huglin, *Polymer* **5**, 135 (1964).
2. T. Fujimoto, H. Narakawa, and M. Nagasawa, *Macromolecules* **3**, 57 (1970).
3. I. Noda, T. Horikawa, T. Kato, T. Fujimoto, and M. Nagasawa, *Macromolecules* **3**, 795 (1970).
4. D. Rahlwes, J. E. L. Roovers, and S. Bywater, *Macromolecules* **10**, 604 (1977).
5. C. Price and D. Woods, *Polymer* **14**, 82 (1973).
6. M. Takaki, R. Asami, and M. Ichikawa, *Macromolecules* **10**, 850 (1977).
7. P. Ho, Y. Yu, C. Lin, and P. Chiang, *CA* **64**, 2184g (1966).
8. G. Greber and J. Tolle, *Makromol. Chem.* **53**, 208 (1962).
9. G. Greber, J. Tolle, and W. Burchard, *Makromol. Chem.* **71**, 47 (1964).
10. G. Greber, *Makromol. Chem.* **101**, 104 (1967).
11. P. Rempp and D. Decker-Freyss, *J. Polym. Sci., Part C* **16**, 4027 (1967).
12. A. Dondos and P. Rempp, *C. R. Hebd. Seances Acad. Sci.* **258**, 4045 (1964).
13. G. Greber and G. Egle, *Makromol. Chem.* **59**, 174 (1963).
13a. G. Greber and G. Egle, *Makromol. Chem.* **62**, 196 (1963).
14. J. E. L. Roovers, *Polymer* **16**, 827 (1975).
15. M. Takaki, R. Asami, and M. Mizuno, *Macromolecules* **10**, 845 (1977).
16. G. Finaz, Y. Gallot, T. Parrod, and P. Rempp, *J. Polym. Sci.* **58**, 1363 (1962).
17. Y. Gallot, P. Rempp, and J. Parrod, *J. Polym. Sci., Part B* **1**, 329 (1963).
18. S. P. Mitzengendler, G. A. Andrews, K. T. Sokolova, and A. A. Korotkov, *Polym. Sci. USSR* (*Engl. Transl.*) **4**, 436 (1963).
19. M. A. Twaik, M. Tahan, and A. Zilkha, *J. Polym. Sci., Polym. Chem. Ed.* **7**, 2469 (1969).
20. D. P. Tate, A. F. Halasa, F. J. Webb, R. W. Koch, and A. E. Oberster, *J. Polym. Sci., Polym. Chem. Ed.* **9**, 139 (1971).
21. J. C. Falk, R. J. Schlott, and D. F. Hoeg, *J. Macromol. Sci., Chem.* **A7**, 1647 (1973).
22. B. W. Brooks, *J. Polym., Sci., Part B* **5**, 641 (1967).

23. J. Heller and D. B. Miller, *J. Polym. Sci., Part B* **7**, 141 (1969).
24. Y. Mimoura and H. Harada, *J. Polym. Sci., Polym. Chem. Ed.* **7**, 3 (1969).
25. J. C. Falk, R. J. Schlott, D. F. Hoeg, and J. F. Pendleton, *Rubber Chem. Technol.* **46**, 1044 (1973).
26. K. Shiina and Y. Minoura, *J. Polym. Sci., Polym. Chem. Ed.* **4**, 1069 (1966).
27. G. Lechermeier, C. Pillot, and T. Gole, *J. Appl. Polym. Sci.* **19**, 1979 (1975).
28. Y. Minoura, H. Hironaka, T. Kasabo, and Y. Ueno, *J. Polym. Sci., Polym. Chem. Ed.* **6**, 2773 (1968).
29. G. Lechermeier, A. Revillon, and C. Pillot, *J. Polym. Sci., Polym. Chem. Ed.* **19**, 1989 (1975).
30. J. C. Falk and R. J. Schlott, *J. Macromol. Sci., Chem.* **A7**, 1663 (1973).
31. B. A. Feit, A. Bar-nun, M. Lahar, and A. Zilkha, *J. Appl. Polym. Sci.* **8**, 1869 (1964).
32. S. Sasson and A. Zilkha, *Eur. Polym. J.* **5**, 315 (1969).
33. A. B. Goshell, J. A. Gervasi, D. K. Woods, and V. Stannett, *J. Polym. Sci., Part C* **22**, 611 (1968).
34. G. Greber and G. Egle, *Makromol. Chem.* **54**, 136 (1962).
35. G. Goutiere and J. Gole, *C. R. Hebd. Seances Acad. Sci.* **257**, 674 (1963); *CA* **59**, 88930; *Bull. Soc. Chim. Fr.*, p. 162 (1965).
36. J. Gole, G. Goutiere, and P. Rempp, *C. R. Hebd. Seances Acad. Sci.* **254**, 3867 (1962); *CA* **57**,10033t.
37. H. Ringsdorff, *J. Polym. Sci., Part C* **4**, 987 (1963).
38. A. Dondos and P. Rempp, *Bull. Soc. Chim. Fr.*, p. 2313 (1962); *CA* **58**, 8052; *C. R. Hebd. Seances Acad. Sci.* **256**, 4443 (1963); *CA* **59**, 6531h.
39. A. Dondos, *Bull. Soc. Chim. Fr.* p. 2762 (1963); *CA* **60**, 10814e.
40. A. J. Chalk and T. J. Hoogeboom, *J. Polym. Sci., Polym. Chem. Ed.* **7**, 1359, 2537 (1969).
41. E. Cohen and A. Zilkha, *J. Polym. Sci., Polym. Chem. Ed.* **7**, 1881 (1969).
42. M. Tahan, B. Yom-Tov, and A. Zilkha, *Eur. Polym. J.* **5**, 499 (1969).
43. M. Tahan and A. Zilkha, *Eur. Polym. J.* **5**, 347 (1969); *J. Polym. Sci., Polym. Chem. Ed.* **7**, 1839 (1969).
44. S. Sasson and A. Zilkha, *Eur. Polym. J.* **5**, 369 (1969).
45. M. Tahan and A. Zilkha, *J. Polym. Sci., Polym. Chem. Ed.* **7**, 1815, 1825 (1969).
46. Y. Avny and A. Zilkha, *Isr. J. Chem.* **3**, 207 (1966).
47. A. S. Bank, *Sin. Vysokomol. Soedin.*, p. 90 (1972); *CA* **79**, 20496a.
48. Y. Avny, B. Yom-Tov, and A. Zilkha, *J. Appl. Polym. Sci.* **9**, 3737 (1965).
49. Y. Avny, S. Migdal, and A. Zilkha, *Eur. Polym. J.* **2**, 355, 367 (1966).
50. J. J. O'Malley and R. H. Marchessault, *J. Polym. Sci.* **B3**, 685 (1965).
51. R. P. Foss, H. W. Jacobson, H. N. Cripps, and W. H. Sharkey, *Macromolecules* **9**, 373 (1976).
52. J. F. Harris, Jr. and W. H. Sharkey, *Macromolecules* **10**, 504 (1977).
53. W. H. Buck, *Rubber Chem. Technol.* **50**, 109 (1977).
54. C. A. Sundet, R. C. Thamm, J. M. Meyer, W. H. Buck, S. W. Caywood, R. M. Subramanian, and B. C. Anderson, *Macromolecules* **9**, 371 (1976).
55. J. R. Schaefgen and P. J. Flory, *J. Am. Chem. Soc.* **70**, 2709 (1948).
56. B. J. Bauer and L. J. Fetters, *Rubber Chem. Technol.* **51**, 406 (1978).
57. M. Morton, T. E. Helminiak, S. D. Gadkary, and F. Bueche, *J. Polym. Sci.* **57**, 471 (1962).
58. T. A. Orofino and F. Wenger, *J. Phys. Chem.* **67**, 566 (1963).
59. S. P. S. Yen, *Makromol. Chem.* **81**, 152 (1965).
60. T. Altares, Jr., D. P. Wyman, V. R. Allen, and K. Meyerson, *J. Polym. Sci., Part A* **3**, 4131 (1965).
60a. J. C. Meunier and R. Van Leemgut, *Makromol. Chem.* **142**, 1 (1971).
61. J. A. Gervasi and A. B. Gosnell, *J. Polym. Sci., Polym. Chem. Ed.* **4**, 1391 (1966).
62. D. Decker and P. Rempp, *C. R. Hebd. Seances Acad. Sci., Ser. C* **261**, 1977 (1965); P. Rempp, *Polym. Prepr., Am. Chem. Soc., Div. Polym. Chem.* **7**, 141 (1966).

63. A. Kohler, J. G. Zilliox, P. Rempp, J. Polacek, and I. Koessler, *Eur. Polym. J.* **8**, 627 (1972).
64. D. J. Worsfold, J. G. Zilliox, and P. Rempp, *Can. J. Chem.* **47**, 3379 (1969).
65. J. G. Zilliox, *Makromol. Chem.* **156**, 121 (1972).
66. T. Masuda, Y. Ohta, and S. Onogi, *Macromolecules* **4**, 763 (1971).
67. J. E. Herz and C. Strazielle, *C. R. Hebd. Seances Acad. Sci., Sec. C* **272**, 747 (1971).
68. J. E. Herz, M. Hert, and C. Strazielle, *Makromol. Chem.* **160**, 213 (1972).
69. M. Hert, C. Strazielle, and J. E. Herz, *C. R. Hebd. Seances Acad. Sci., Ser. C* **276**, 395 (1973).
70. C. Strazielle and J. E. Herz, *Eur. Polym. J.* **13**, 223 (1977).
71. J. E. L. Roovers and S. Bywater, *Macromolecules* **5**, 384 (1972).
72. J. E. L. Roovers and S. Bywater, *Macromolecules* **7**, 443 (1974).
73. W. A. J. Bryce, G. McGibbon, and J. E. Meldrum, *Polymer* **11**, 394 (1971).
74. H. Eschwey and W. Burchard, *Polymer* **16**, 180 (1975).
75. R. P. Zelinski and C. F. Wofford, *J. Polym. Sci., Part A* **3**, 93 (1965).
76. B. L. Johnson, H. E. Adams, F. C. Weissert, and K. Farhat, *Proc. Int. Rubber Conf., 5th, 1967*, p. 29 (1968).
77. C. A. Uraneck and J. N. Short, *J. Appl. Polym. Sci.* **14**, 1421 (1970).
78. A. Ghijsels and H. J. M. A. Mieras, *J. Polym. Sci., Polym. Phys. Ed.* **11**, 1849 (1973).
79. H. H. Meyer and W. Ring, *Kautsch. Gummi, Kunstst.* **10**, 526 (1971).
80. K. Osaki, Y. Mitsuda, R. M. Johnson, J. L. Schrag, and J. D. Ferry, *Macromolecules* **5**, 17 (1972).
81. Y. Tanaka, K. Matsuzaka, and N. Yamahara, Japanese Patent 71/18,492 (1971); *CA* **75**, 99159g; 71/18,494; *CA* **75**, 99160k.
82. L. J. Fetters and M. Morton, *Macromolecules* **7**, 552 (1974).
83. N. Hadjichristidis and J. E. L. Roovers, *J. Polym. Sci., Polym. Phys. Ed.* **12**, 2521 (1974).
84. N. Hadjichristidis, A. Guyot, and L. J. Fetters, *Macromolecules* **11**, 889 (1978).
85. N. Hadjichristidis and L. J. Fetters, *Macromolecules* **13**, 191 (1980).
86. G. Quack, L. J. Fetters, N. Hadjichristidis, and R. N. Young, *Ind. Eng. Chem. Prod. Res. Dev.* **19**, 507 (1980).
87. E. E. Bostick, U. S. Patent 3,378,521 (1968); *CA* **69**, 3464q (1968).

Chapter

11 α,ω-Difunctional Polymers

As indicated previously, the nonterminating character of homogeneous anionic polymerization also enables the synthesis of polymers having terminal functional groups, by use of a suitable additive at the completion of the polymerization reaction. This possibility, together with the narrow molecular weight distribution attainable with these systems, provides another powerful synthetic tool for the design of polymers with a predictable molecular structure.

The greatest interest, of course, has been in such macromolecules having a useful functional group at *each end*, i.e., the α,ω-difunctional polymers. This is because such polymers can be used, together with suitable linking agents, to carry out two important operations, i.e., (i) chain extension of short chains to long ones by means of difunctional linking agents, and (ii) formation of networks by use of polyfunctional linking agents. Both of these are of great industrial interest, since they are the basis of "liquid polymer" technology whereby polymeric materials may be cast or injection molded, exemplified by the urethanes or the recently developed "reaction injection molding" (RIM) systems. The greatest interest has probably been shown by the rubber industry, since rubber technology is based on the formation of networks (vulcanization) by the crosslinking of long chains, and this involves an energy-intensive mixing operation of high-viscosity materials (along with some undesirable chain degradation reactions caused by the high temperatures and shearing forces). The use of liquid precursors which can be end-linked into the desired network by polyfunctional agents could offer not only processing advantages but perhaps superior performance properties of the final network as well.

Since the 1,3-dienes are known to yield elastomeric polymers, it is not surprising to find that butadiene, for example, has been used extensively for the synthesis of difunctional, short-chain polymers. The synthesis of α,ω-difunctional polydienes has been reviewed recently.[1] A number of publications have described the synthesis of such polymers either by free radical[1-4] or anionic[1,5-7a] methods. The free radical methods suffer from a broad distribution of molecular weights and from branching and other side

TABLE 11.1 α,ω-Difunctional Polystyrene[a]

Additive	Functional group	References
Ethylene oxide	$-OH$	16–18
Styrene epoxide	$-OH$	19
Carbon dioxide[b]	$-COOH$	20–22
Esters and anhydrides		
Ethyl benzoate	$-COC_6H_5$	23
Ethyl *p*-Aminobenzoate	$-COC_6H_4NH_2$	24
Ethyl adipate	$-COOC_2H_5$	25
Ethyl sebacate	$-COOC_2H_5$	25
Succinic anhydride	$-CO(CH_2)_2COOH$	26
Phthalic anhydride	$-COC_6H_4COOH$	26
Halides		
Bromine	$-Br$	27, 28
Iodine	$-I$	27
Phosphorus tribromide	$-PBr_2$	25
Phosgene	$-COCl$	25, 29[c]
Thionyl chloride	$-SOCl$	25
t-Butyl bromide	$-C(CH_3)_3$	27
t-Butyl iodide	$-C(CH_3)_3$	27
Dibromobutane	$-(CH_2)_4Br$	25
Dibromohexane	$-(CH_2)_6Br$	25
Allyl chloride	$-CH{=}CH_2$	26
Allyl bromide	$-CH{=}CH_2$	26
Bromobenzene	$-C_6H_5$	26
Benzoyl chloride	$-COC_6H_5$	26, 27
Benzyl chloride	$-CH_2C_6H_5$	26
Benzyl bromide	$-CH_2C_6H_5$	26
p-Bromoaniline	$-C_6H_4NH_2$	26
N-Bromophthalimide	−phthalimide	26
Bromonaphthalene	$-C_{10}H_8$	27
Dimethylchlorosilane	$-Si(H)(CH_3)_2$	30

[a] Initiator: Na naphthalene. Solvent: H_4-furan.
[b] Initiator: Na-α-methylstyrene.
[c] Initiator: Na-α-methylstyrene. Solvent: toluene/H_4-furan.

TABLE 11.2 α,ω-Difunctional Polybutadiene

Functional group	Additive	Initiator	Solvent	References
–OH	Ethylene oxide	Li naphthalene	H_4-furan	20, 31
	Ethylene oxide	Na or Li	$C_2H_5OC_2H_5$	32
	Ethylene oxide	Na naphthalene	H_4-furan	33
	Ethylene oxide	Li 2-methyl-naphthalene	Hexane	34
	Ethylene oxide	Dilithiopoly-isoprene	Hydrocarbon	35
	Propylene oxide	Na or Li	$C_2H_5OC_2H_5$	32
	Styrene oxide	Na or Li	$C_2H_5OC_2H_5$	32
	Aldehydes, ketones	Na naphthalene	H_4-furan	33
	Formaldehyde	Li naphthalene	H_4-furan	20
	Halohydrins	Na naphthalene	H_4-furan	33
	Glycidaldehyde	1,2-Dilithio-1,2-diphenylethane	Hydrocarbons	36
	1,2,3,4-Diepoxy-butane	1,2-Dilithio-1,2-diphenylethane	Hydrocarbons	36
	Sulfur dioxide	1,2-Dilithio-1,2-diphenylethane	Hydrocarbons	37
	Oxygen	1,2-Dilithio-1,2-diphenylethane	Hydrocarbons	37
–SH	Carbon disulfide	Na naphthalene	H_4-furan	33
	Alkylene sulfides	Na naphthalene	H_4-furan	33
	Cyclic disulfides	1,2-Dilithio-1,2-diphenylethane	Toluene	38
	Sulfur	1,2-Dilithio-1,2-diphenylethane	Hydrocarbons	37
	Sulfur	Na naphthalene	H_4-furan	39
	Sulfur	Na α-methylstyrene	H_4-furan	39
–COOH	Carbon dioxide	?	?	5
	Carbon dioxide	Lithium	H_4-furan	40
	Carbon dioxide	1,2-Dilithio-1,2-diphenylethane	Cyclohexane	41, 42
	Carbon dioxide	1,2-Dilithio-1,2-diphenylethane	Hydrocarbons	43
	Carbon dioxide	Na naphthalene	H_4-furan	33
	Carbon dioxide	Na α-methylstyrene	H_4-furan	39
	Carbon dioxide	Na α-methylstyrene	H_4-furan	44
	Carbon dioxide	Li naphthalene	H_4-furan	20
	Carbon dioxide	Aryl ketone dilithium	Toluene	45
	Carbon dioxide	Dialkyl *m*-xylylene dilithium	Cyclohexane	46
–CSSH	Carbon disulfide	Na naphthalene,	H_4-furan	39
		Na α-methylstyrene	H_4-furan	39
$-SO_3H$	Sulfuryl chloride	1,2-Dilithio-1,2-diphenylethane	Hydrocarbons	37, 43

(*table continues*)

TABLE 11.2 *(continued)*

Functional group	Additive	Initiator	Solvent	References
$-Cl$	Chlorine	1,2-Dilithio-1,2-diphenylethane	Hydrocarbons	37
	Chlorine	Na naphthalene or Na α-methylstyrene	H_4-furan	39
$-Br$	Bromine	Na naphthalene or Na α-methylstyrene	H_4-furan	39
$-OCH_2Cl$	Bischloromethyl ether	1,2-Dilithio-1,2-diphenylethane	Cyclohexane	47
$-OCH_2Br$	Bisbromomethyl ether	1,2-Dilithio-1,2-diphenylethane	Cyclohexane	47
$-COCH_3$	Acetyl chloride	1,2-Dilithio-1,2-diphenylethane	Hydrocarbons	37
$-CN$	Cyanogen chloride	1,2-Dilithio-1,2-diphenylethane	Hydrocarbons	37
$-C_6H_4NO_2$	*p*-Chloronitrobenzene	1,2-Dilithio-1,2-diphenylethane	Hydrocarbons	37
$-NH_2$	Alkyleneimines	Na naphthalene	H_4-furan	33
$-P(C_2H_5)_2$	$(C_2H_5)_2PCl$	1,4-Dilithio-1,1,4,4-tetraphenylbutane	C_6H_{12}/TMEDA	48
$-\underset{\diagdown O \diagup}{H_2C-CH_2}$	Epichlorohydrin	Na α-methylstyrene or Na naphthalene	H_4-furan	39

reactions[8,9] which lead to functionalities greater or less than two.[10,11] As indicated above, the homogeneous anionic systems offer a more elegant method for the preparation of difunctional polymers. However, in the case of the 1,3-dienes used for the synthesis of elastomers, it is necessary to carry out the polymerization in the absence of polar solvents, e.g., ethers, which favor the formation of a high proportion of side-vinyl groups in the polymer (see Chapter 7). Hence this requires the use of a difunctional initiator, e.g., an organodilithium species, which must be soluble in hydrocarbon media, and these have been very difficult to prepare. A recent publication reviews the problems involved in the preparation of such organodilithium initiators.[7a]

It is only recently that success has been achieved in the preparation of efficient organodilithium initiators[7,12–14] which can be prepared in high concentrations (0.5–1.0*M*) in nonpolar solvents. Some of these have been used to prepare a series of α,ω-dihydroxypolyisoprenes[7] having a high 1,4 content and a very narrow molecular weight distribution, and these have been successfully linked by means of a triisocyanate into an elastic network of

TABLE 11.3 α,ω-Difunctional Polyisoprene

Functional group	Additive	Initiator	Solvent	References
–OH	Ethylene oxide	Li naphthalene	H_4-furan	31
	Ethylene oxide	Li naphthalene	H_4-furan	49[a]
	Ethylene oxide	Li α-methyl-naphthalene	Hexane	39
	Ethylene oxide	α,ω-Dilithio-oligo-2,4-hexadiene	$C_6H_6/(C_2H_5)_3N$	7
	Formaldehyde	Li naphthalene	H_4-furan	49[a]
–SH	Cyclic disulfides	1,2-Dilithio-1,2-diphenylethane	Toluene	38
–COOH	Carbon dioxide	1,2-Dilithio-1,2-diphenylethane	Cyclohexane	41, 42
	Carbon dioxide	Li naphthalene	H_4-furan	49[a]
	Carbon dioxide	Dialkyl *m*-xylylene-dilithium	Cyclohexane	46
$-SO_3H$	Sulfuryl chloride	1,2-Dilithio-1,2-diphenylethane	Hydrocarbons	43
$-\underset{\textstyle OH}{\underset{\textstyle \vert}{C}}HC_6H_4N(CH_3)_2$	4-Dimethylamino-benzaldehyde	1,2-Dilithio-1,2-diphenylethane	Cyclohexane	50[b]

[a] Also includes polystyrene and polybutadiene.
[b] Also includes polybutadiene.

controlled uniformity.[15] The properties of such networks have been recently discussed.[15]

To date the interest in preparing difunctional polymers by anionic polymerization has been almost exclusively limited to the more common, commercially important monomers which are amenable to these polymerizations, i.e., styrene, butadiene, and isoprene. A listing of the various attempts to prepare *difunctional* polymers from these monomers is contained in Tables 11.1, 11.2, and 11.3. This information was mainly obtained from the literature and, with the exception of any work reported from our laboratories, is listed without any critical evaluation. The reader will, therefore, have to exercise the usual caution in accepting these results, in view of the known sensitivity of anionic polymerization to impurities and side reactions.

References

1. D. M. French, *Rubber Chem. Technol.* **42**, 71 (1969).
2. S. F. Reed, *J. Polym. Sci., Polym. Chem. Ed.* **9**, 2029, 2147 (1971); **10**, 649, 2025, 2493 (1972).

3. C. H. Bamford, A. D. Jenkins, and R. P. Wayne, *Trans. Faraday Soc.* **56**, 932 (1960).
4. W. H. Stubbs, C. R. Gore, and C. S. Marvel, *J. Polym. Sci., Polym. Chem. Ed.* **4**, 1898 (1966).
5. C. A. Uranek, H. L. Hsieh, and O. G. Buck, *J. Polym. Sci.* **46**, 535 (1960).
6. S. F. Reed, *J. Polym. Sci., Polym. Chem. Ed.* **10**, 1187 (1972).
7. M. Morton, L. J. Fetters, J. Inomata, D. C. Rubio, and R. N. Young, *Rubber Chem. Technol.* **49**, 303 (1976).
7a. L. J. Fetters, C. W. Kamienski, R. C. Morrison, and R. N. Young, *Macromolecules* **12**, 344 (1979).
8. R. D. Law, *J. Polym. Sci., Polym. Chem. Ed.* **7**, 2097 (1969); **9**, 539 (1969); **11**, 175 (1973).
9. J. P. Consaga, *J. Appl. Polym. Sci.* **14**, 2157 (1970).
10. J. N. Anderson, S. K. Baczek, H. E. Adams, and L. E. Vescelius, *J. Appl. Polym. Sci.* **19**, 2255 (1975).
11. S. K. Baczek, J. N. Anderson, and H. E. Adams, *J. Appl. Polym. Sci.* **19**, 2269 (1975).
12. M. Morton and L. J. Fetters, U. S. Patent 3,663,634 (1972).
13. L. J. Fetters and M. Morton, *Macromolecules* **2**, 453 (1969).
14. L. J. Fetters, U. S. Patent 3,848,008 (1974).
15. M. Morton and D. C. Rubio, *Plast. Rubber: Mater. Appl.* **3**, 139 (1978).
16. H. Brody, D. H. Richards, and M. Szwarc, *Chem. Ind. (London)* **45**, 1473 (1958).
17. D. H. Richards and M. Szwarc, *Trans. Faraday Soc.* **55**, 1644 (1959).
18. E. Franta and P. Rempp, *C. R. Hebd. Seances Acad. Sci.* **254**, 674 (1962).
19. Y. Shimura and W. S. Lin, *J. Polym. Sci., Polym. Chem. Ed.* **8**, 2171 (1970).
20. Y. Yamashita and T. Hane, *J. Polym. Sci., Polym. Chem. Ed.* **11**, 425 (1973).
21. Y. Yamashita, Y. Nakamura, and S. Kojima, *J. Polym. Sci., Polym. Chem. Ed.* **11**, 823 (1973).
22. Y. Yamashita, Japan Kokai 73/00,792 (1973); *CA* **78**, 111979f.
23. P. Rempp, V. I. Volkov, J. Parrod, and C. Sadron, *Bull. Soc. Chim. Fr.* p. 919 (1960).
24. M. H. Loucheux, G. Mayer, and P. Rempp, *C. R. Hebd. Seances Acad. Sci.* **252**, 2552 (1961).
25. G. Finaz, Y. Gallot, J. Parrod, and P. Rempp, *J. Polym. Sci.* **58**, 1363 (1962).
26. P. Rempp and M. H. Loucheux, *Bull. Soc. Chim. Fr.* p. 1457 (1958).
27. D. H. Richards and D. A. Salter, private communication.
28. F. T. Burgess, A. V. Cunliffe, T. R. MacCallum, and D. H. Richards, *Polymer* **18**, 719 (1977).
29. M. N. Berger, J. J. K. Boulton, and B. W. Brooks, *J. Polym. Sci., Polym. Chem. Ed.* **7**, 1339 (1969).
30. G. Greber, E. Reese, and A. Balciunas, *Farbe Lack* **70**, 249 (1964); *CA* **61**, 5800a.
31. E. J. Goldberg, U. S. Patent 3,055,952 (1962).
32. National Distillers and Chemical Corp., British Patent 946,092 (1964).
33. B. F. Goodrich Co., British Patent 964,259 (1964).
34. K. Suzuki, O. Oomura, H. Kahi, Y. Kitamura, and Y. Yoshimasa, Japanese Patent 72/06,724 (1972); *CA* **77**, 21179v.
35. S. F. Reed, *J. Polym. Sci., Polym. Chem. Ed.* **10**, 1187 (1972).
36. Phillips Petroleum Co., British Patent 945,851 (1964).
37. Phillips Petroleum Co., British Patent 906,315 (1962).
38. J. W. Cleary, U. S. Patent 3,048,568 (1962).
39. K. S. Dennis, U. S. Patent 3,458,491 (1969); *CA* **71**, 71394h.
40. E. Schoenberg, *J. Polym. Sci.* **49**, 59 (1961).
41. R. P. Zelinski and C. W. Strobel, U. S. Patent 3,108,994 (1963).
42. Phillips Petroleum Co., British Patent 921,803 (1963).
43. W. B. Reynolds, U. S. Patent 3,074,917 (1963).
44. Dow Chemical Co., French Patent 1,536,112 (1968).
45. H. L. Hsieh and W. J. Trepka, U. S. Patent 3,410,836 (1968); *CA* **70**, 29568x.
46. R. Foss, Ger. Offen. 2,307,740 (1973); *CA* **80**, 97171u.

47. R. P. Zelinski and H. L. Hsieh, U. S. Patent 3,078,254 (1963).
48. H. E. De la Mare and F. E. Neuman, Ger. Offen. 1,938,289 (1970); *CA* **72**, 90999w.
49. K. Hayashi and C. S. Marvel, *J. Polym. Sci., Part A* **2**, 2571 (1964).
50. R. P. Zelinski, H. L. Hsieh, and C. W. Strobel, U. S. Patent 3,109,871 (1963).

Index